TECHNICAL EDITING

Edited by

B. H. WEIL

Chief Editor, Technical Information Division
Esso Research and Engineering Company
Linden, New Jersey

CONTRIBUTING AUTHORS

Gordon H. Bixler
Allen D. Bliss
William F. Bland
David N. Collins
Lucie D. Connolly
Ethaline Cortelyou
Burr J. French
G. G. Hawley
E. H. Huntsman
I. H. Jenks
B. A. Jones
John C. Lane
Louis Meites
Mary Ann Mento
Samuel A. Miles
P. M. Reyling
Morris D. Schoengold
T. E. R. Singer
Julian F. Smith
Irlene Roemer Stephens

REINHOLD PUBLISHING CORPORATION
NEW YORK

CHAPMAN & HALL, LTD., LONDON

Library of Congress Catalog Card Number: 58-10426

REINHOLD PUBLISHING CORPORATION
Publishers of Chemical Engineering Catalog, Chemical Materials Catalog, "Automatic Control," "Materials in Design Engineering," "Progressive Architecture"; Advertising Management of the American Chemical Society

Printed in the U.S.A.

Foreword

Seldom has a book been issued that filled a need at just the right time as does "Technical Editing." This is not a coincidence—the Editor and his colleagues set out well in advance to gather the information they wanted to publish, to present it at an ACS Symposium, and then to organize it for this book.

Their planning was purposeful. The contributing authors, all leaders in their fields, realized that their experience should be shared by others to help meet the rapidly growing problems in technical communication—problems internal to government, educational, and industrial organizations; problems internal to professional societies and trade groups; and, even more important, problems of communicating technical messages to a public which has an aroused interest in and concern with science.

The organization and content of the book will contribute to the realization by managements of government, educational, and industrial organizations of the scope and skills of technical editors and of their growing importance as the link between scientific findings and their applications.

The pleasure of writing a Foreword for this book is enhanced by my sincere appreciation of its timely contribution to the field of technical communication.

C. E. STARR, Jr.
Manager of Technical and Public Information
Esso Research and Engineering Company

Preface

During the past 20 years, great emphasis has been placed on technical reporting in industry and government. Scientists, engineers, supervisors, and top management have come to realize the importance of clear and adequate transmission of technical messages—the need for properly reporting the findings of billions of dollars worth of research and engineering. Many books and manuals have appeared on how to write, how to write technical papers, how to write technical reports, how to prepare slides, how to present papers and oral reports, and how to conduct conferences. Courses in technical writing have been given in colleges and industry, and by institutes and consultants. Staff writers have been employed by many organizations to prepare technical documents on a professional basis. Divisions of scientific societies have been organized to promote better technical communication. A Society of Technical Writers and Editors has come into being.

Most of these efforts have logically concentrated on improving at-the-source preparation of technical documents, on training the scientist and engineer to handle his own technical communicating properly. But along with these endeavors—as instigators as well as guides—has come a new (or at least vastly expanded) group of technical professionals known as "technical editors."

Technical editing, as such, is far from new. Technical books and journals have always required technical editors to approve the pertinency of manuscripts, maintain publication standards, and actually publish these documents. Similarly, project supervisors have always had to approve the reports and papers of their staff members—an approval which often calls for editing of the highest order.

What is novel or significant in recent developments, however, is the emergence of a group of technical men and women who dedicate their full-time efforts to improving "internal" technical communication and communicating media, and the communicators with whom they work. These technical editors usually work side by side with the project scientist or engineer. They set publication standards for him, provide him with writing guides, review his documents for proper presentation, counsel with him on problems, and handle the publication of his documents.

The role of this new type of technical *editor*, of course, is not always distinct, either functionally or individually, from that of the staff *writer*. In small organizations, one man may do staff writing as well as technical editing; in large organizations, technical editors and writers are often members of the same group. Their close association is also recognized in their scientific society, the "Society of Technical Writers *and Editors*."

Until now, however, attention has been focused largely on the activities and functions of technical writers. The role of the technical editor has been dealt with only in an occasional paragraph or paper. Even a few conferences on "editorial problems" have not made the focus sharp, and have not covered the full breadth of the field.

For this reason—to present a clear picture of most phases of technical editing and of most types of technical editors—the editor of this book, in 1956, envisioned and planned a "Symposium on Technical Editing." With the help and advice of 20 colleagues in this field, he served as chairman of this symposium in New York City on September 9, 1957, at the meeting of the Division of Chemical Literature of the American Chemical Society. The papers there presented, released for this purpose by the American Chemical Society, form the nucleus for this book.

This book is intended for the experienced technical editor, to serve as a check against his own practices and attitudes. It is intended for the management of technical industries and agencies, as an indication of what technical editors can do and how they should be directed. It is intended for the fledgling editor, who all too often is given responsible assignments for which he is not yet psychologically prepared or sufficiently experienced. Finally, it is intended for the technical student and the student of technical journalism or technical communication who is attracted to editing as a profession.

In essence, the book is a review of the field of technical editing from the "what," "why," and "how" standpoints, with a touch of "when" and "where." "What" is edited falls into the categories of papers, reports, slides, tables, figures, journals, business magazines, books, manuals, abstracts, and translations, to mention but a few. "Why" they are edited is clear—to maximize their communication impact. "How" they are edited to accomplish this varies with the medium, and ranges from "structuring" technical articles to deleting extraneous data from slides. "When" and "Where" are concerned with timeliness and timing.

"Technical editing," in this book, has usually been interpreted broadly to include "technical publishing," since the technical editor is almost always a publisher as well as "simply" the wielder of a blue pencil. He is also a teacher, with a bit of artist thrown in.

The divisions of this book have been chosen arbitrarily to relate chapters dealing with specific phases of technical editing and/or the editing of specific types of technical documents and graphic aids. These divisions cannot be defended on the basis of pure logic, however, for the unit concepts and aspects are too interwoven. The internal editing of technical papers, for example, is a phase of the editing (publishing) of technical journals. The editing of lantern slides and other graphic aids is normally internal editing, but has been discussed in the division of the book that also deals with the editing of illustrations and tables for reports and papers (also related to the editing of journals). Enough has been said, I think, to stress the importance of referring to several chapters for phases of a given subject, despite every attempt at unity.

The student or practioner of any phase of technical editing will do well, also, to use Chapter Seventeen extensively—"Editing Know-How—Techniques and Tools." By agreement among the authors of this book, coverage of the techniques and tools of technical editing has been assigned to the authors of this comprehensive review and guide to further information.

The authors of chapters in this book are men and women who are long experienced in working with people as well as with words and media. That is important, for human dignity is well worth protecting in an age in which the rush of events tends to obscure the role of the individual. Nowhere in this book will you find dictatorial statements. But you will find many instances of the understanding of human traits as they apply to communicating, and many suggestions as to how editors can inspire their colleagues to produce their best work.

As editor of this book, it has been my privilege to work cooperatively with the authors of chapters—to edit editors, as it were. In this process we have attempted to turn meeting papers into documents of more permanent value. We have scrupulously avoided changing approach and style, however. Some chapters say "we," "you," and "I" with disarming freshness. Others are more formal without being stiff. Working with experienced editors as writers has been a pleasure which few book editors have had a chance to experience.

Acknowledgments for advice and assistance are due to so many people that I am forced to mention but a few: T. E. R.

Singer, for encouragement; Walter Murphy of the American Chemical Society, for his kindness in suggesting certain authors (editors) and for permission to publish the ACS papers on which many chapters are based; Mary Carey, for skillful secretarial assistance; and my wife, Carolyn, for patiently waiting out the period of planning, writing, and editing, and for help with the proofreading. Thanks are also due to C. E. Starr, Jr., for his courtesy in writing the Foreword.

As far as I know, this is the first book on technical editing. Suggestions for future editions are sincerely invited, for much remains to be written on this subject.

B. H. WEIL

Linden, New Jersey
May 1, 1958

Contents

FOREWORD iii

PREFACE v

Chapter One

PSYCHOLOGICAL, EDUCATIONAL, AND PROFESSIONAL ASPECTS OF TECHNICAL EDITING. *B. H. Weil* 1

PART I. EDITING INTERNAL DOCUMENTS

Chapter Two

EDITING TECHNICAL REPORTS. *Irlene Roemer Stephens* and *P. M. Reyling* 19

Chapter Three

INTERNAL EDITING OF TECHNICAL PAPERS AND ARTICLES. *Burr J. French* 31

Chapter Four

EDITING ORGANIZATION-SPONSORED MAGAZINES AND NEWSLETTERS. *John C. Lane* 41

Chapter Five

EDITING FROM THE PATENT STANDPOINT 51

Chapter Six

EDITING AND PUBLISHING TECHNICAL ABSTRACTS FOR INTERNAL USE. *Morris D. Schoengold, B. H. Weil,* and *Mary Ann Mento* 62

Chapter Seven

TRAINING THE INTERNAL TECHNICAL EDITOR. *Lucie D. Connolly* 80

PART II. EDITING JOURNALS

Chapter Eight

EDITING OF TECHNICAL-SOCIETY JOURNALS: THE EDITOR'S VIEWPOINT. *Allen D. Bliss* 91

Chapter Nine

TECHNICAL EDITING: THE REVIEWER'S VIEWPOINT. *Louis Meites* 103

Chapter Ten

EDITING THE TECHNICAL BUSINESS MAGAZINE. *William F. Bland* 115

Chapter Eleven

EDITING TECHNICAL NEWS RELEASES. *Gordon H. Bixler* 133

PART III. EDITING BOOKS AND MANUALS

Chapter Twelve

PUBLISHER EDITING OF TECHNICAL BOOKS. *G. G. Hawley* 155

Chapter Thirteen

EDITING TECHNICAL-EQUIPMENT MANUALS. *Samuel A. Miles* 167

PART IV. EDITING GRAPHIC AIDS AND OTHER EXHIBITS

Chapter Fourteen

EDITING SLIDES AND OTHER GRAPHIC AIDS. *David N. Collins* and *B. A. Jones* 187

Chapter Fifteen
EDITING ILLUSTRATIONS FOR TECHNICAL REPORTS AND PAPERS. *Ethaline Cortelyou* and *B. A. Jones* 203

Chapter Sixteen
EDITING TABULAR DATA. *Ethaline Cortelyou* 217

PART V. GENERAL

Chapter Seventeen
EDITING KNOW-HOW—TECHNIQUES AND TOOLS. *I. H. Jenks* and *E. H. Huntsman* 233

Chapter Eighteen
EDITING TRANSLATIONS. *Julian F. Smith* and *T. E. R. Singer* 251

Chapter Nineteen
OTHER ASPECTS. *B. H. Weil* 257

AUTHOR INDEX 267

SUBJECT INDEX 271

Chapter One

Psychological, Educational, and Professional Aspects

B. H. WEIL

Chief Editor, Technical Information Division
Esso Research and Engineering Company
Linden, New Jersey

In these days of increased emphasis on technical communication, with technical writers and technical editors employed in many branches of industry and government, there are times when the stress placed on application obscures the basic aspects involved. Technical writers are often so busy doing technical writing that they forget that what they are doing is *writing*, with the technical phases as subject matter. Technical editors are sometimes so overworked that they neglect the fundamentals of *editing*, itself a profession of long standing.

Writing and editing, of course, are allied fields. Writers produce the media which editors edit. The good writer must know the tricks of his trade—semantics and syntax—what to say, and how to say it. The good editor must know these and more—how to review from the standpoint of the ultimate readers, how to alter documents in harmony with their writers, and how to complete the publication of documents in terms of the proper graphic arts.

In modern industry and in government, of course, technical writers must sometimes do editing, and technical editors must often do writing. This is fortunate, for it builds an appreciation of the differences involved as well as experience in the related field. A good editor really ought to be a good writer who prefers to edit because of his interest in working with people as well as words.

PSYCHOLOGICAL ASPECTS

As Regards Writers

Inevitably, the good technical editor must become a student of applied psychology. If he does not know it when he first begins to alter a writer's words, paragraphs, and sections, or proposes that the writer himself should make certain changes, he quickly finds out that he may be tampering with the writer's ego, with explosive results. Since his editing is perforce done in terms of the ultimate reader, and at the direction of management, this means that the editor must develop a good personal relationship with the writers whose documents he edits; such a relationship enables both writer and editor to perform their work as a team.

Such a relationship can be developed only when the technical editor observes several rules:

(1) Editors, writers, and management must agree in advance as to the rules of the game, as expressed either formally in style manuals or less completely by management statement, book contract, or the known policy of journals or societies.

(2) Technical editing must generally be restricted, in so

far as humanly possible, to the indication of or correction of errors in fact or grammar, and to improvement of clarity, consistency, and brevity. The technical editor must not seek to impose his own style of writing or personal preference for words, although he may have to enforce a style established by his organization.

(3) The reason for an editorial change or suggestion must be explained if it is not self-evident—by marginal annotation, in an accompanying note, or in a personal conference. The latter is the preferable means, since a good editor can best project the "helpful" aspects of his work face to face with his writing colleague. Confronted with the need to explain his editing, the technical editor will hesitate to be arbitrary or dictatorial.

(4) The technical editor must really be or become an expert in what he is doing—technical editing—in order to gain and keep the confidence and respect of the writers whose work he edits. He must be an authority in grammar, in technical terminology and abbreviations, in the science of communication as it is now developing, in applied graphic arts, and in the management philosophy which lies behind his work. He must be always available and sincerely willing to help or advise a writer on some fine point at any stage of his work.

(5) The technical editor must feel and express a sincere interest in the writer's work—its subject, the findings reported, and (if possible) the author's personal role. This interest must be reflected in the letter and spirit of the editing required, and in all contacts with the author, written or personal. This is vital indeed in the case of the internal editor, and a matter of good public relations for the book or journal editor, even when he must reject a document.

(6) The technical editor must project his personality as that of a warm, helpful colleague. If he is more informed than the writer in the subject of the document, and the writer knows this, he must avoid any appearance of

superiority and must place stress in his editing on clarity and impact. If he is less informed—as most editors are, since they must usually edit a broad spectrum of subjects—he may profess this and place stress on assisting the writer to put across his message with optimum effectiveness.

The technical editor must be patient in the face of impatience, and sympathetic (but without sycophancy) to rebellion against rule—and yet gain his point where gain it he must. Similarly, he must know when to yield on minor points even when he is right, in order to win more important battles later. He must be willing to listen as a friend to unrelated gripes, in order to be listened to as a friend on related aspects. Overloaded with work or not, he must never appear impatient with the writer, in interview or correspondence.

As Regards Management

If his work with technical writers is to be effective, of course, the technical editor must establish a firm relationship with management, whether this management is that of an industry, publishing house, or journal. While his work is with authors, and the results will be judged by the readers, the technical editor is usually in a real sense a management representative charged with a specific task—to see that technical data are reported in a manner suited to management's goals for them.

This means that the technical editor must clearly understand what management is trying to achieve and that he must keep abreast of changes in goals and philosophy. It means, also, that he must keep management informed of his own techniques, and must clear with management any rules which he may codify.

As Regards Illustrators and Publishers

Since the technical editor usually works closely with illustrators and publishers—by these or such names as draftsmen, industrial artists, photographers, print shops, printers, and the like—he must also develop empathy for them. He must understand not only what they can do for him, in terms of his needs, but also the approaches which will gain the best results. He must learn, for example, that the artist and photographer can often give him superior work if he does not dictate the weight of every line and the placing of the camera for every shot. He must learn the individual characteristics of each man—to some he can supply sketches; from others he will do better to ask for sketches and choose from among them.

The technical editor, except for the editors of books and manuals, is not usually in a position where he can *command* the scheduling of preparation for the graphic materials he requires and the printing of the documents he is publishing. Usually, then, he must walk the fine tightrope called "expediting." He must plan his work to make allowance for the rules of graphics departments, and for their foibles. He must know how often he can ask for special service and have a right to expect it without damage to basic relationships.

Above all, the technical editor must remember that draftsmen, artists, photographers, and printers are craftsmen who take pride in their work, just as he does in his. He must be free and honest with his praise, and careful with his criticism. If he takes pains to treat graphic-arts experts as talented fellow workers, he will be amply rewarded.

As Regards Readers

The efforts of technical writers, technical editors, illustrators, and publishers are all ultimately directed toward producing documents suited to the intended readers. This means, as mentioned earlier, that the technical editor must keep the needs and interests of the intended readers foremost in mind when he edits each document, and that he must judge and mold the writer's efforts in these regards.

Accordingly, the technical editor must remember that readers, being busy people, are inordinately impatient with over-detailed discussions, vague statements, and overuse of synonyms, unfamiliar jargon, and obscure units. He must also strive to produce documents that are as attractively presented as budgets will permit, since he is aware that many readers react unfavorably to reports, manuals, and books that contain long lines of small type (or typing), are poorly duplicated, and are inadequately bound.

If the internal editor has control over distribution procedures, he must take pains not to send too many and too detailed reports to higher echelons of management. To avoid this he may have to turn writer and prepare special summary reports, or he must see that the report writers prepare adequate summaries and streamline their presentation of details. These and other psychological aspects of writers, readers, and editors have been discussed in a recent paper.[7]

EDUCATIONAL ASPECTS

As has been intimated, the technical editor, over and above his monitorial role in improving specific documents, owes it to himself and his employer to help authors to improve their style and correct certain deficiencies (directly or indirectly), and thus to aid in employee development. He can do this most effectively by preparing helpful style manuals, by conducting internal courses in technical writing, and—best of all—by concentrating on proper day-to-day relationships.

Style Manuals

Style manuals, of course, can serve several purposes. They codify the rules of document organization and style which are desired by management, book publisher, or journal. They clarify the reason for such rules, and explain how adherence can render easier the task of writing. They often contain helpful "know-how" hints on how to collect and organize data prior to writing, how to begin writing, how to organize tables, how to prepare figures and photographs, how to prepare footnotes and bibliographies, etc. They may contain lists of abbreviations and symbols. They usually deal with the proper use of parts of speech and punctuation marks.[6]

The preparation of style manuals for internal use is usually the job of the technical editor, since he is in the best position to know what rules and information are needed by the technical writers with whom he works. While he realizes that no style manual will do the full job of educating the technical writer, the editor usually welcomes the opportunity to reduce to writing most of the helpful information in his possession. This then gives him a handbook to which he can refer authors when many questions arise, and a textbook on which he can base formal or informal lectures on technical writing.

The publishers of technical books and journals also save much time and money by making style manuals available to prospective authors. Editing of technical documents for message and presentation is difficult enough without also having each author free to pursue a style of his own—not that some won't anyway.

Internal Courses on Writing

More and more organizations are realizing that their new technical employees—college graduates though they may be—need additional training in technical writing if their abilities in other areas are to be put to optimum use, and if their progress is to be expedited. They have also realized that many of their older employees are equally or even more in need of help to improve their writing. To remedy the problem of poor technical writing quickly, and to monitor quality-control programs, they are employing technical editors. Where they have such editors, they often ask these men and women to plan, conduct, and follow up on courses on technical writing in order to improve the writing abilities of their technical staffs. Or the editor himself sooner or later offers to conduct such a course, or at least to organize one.

The proper conduct and subject matter of internal courses in technical writing have been dealt with in several recent papers,[2, 4, 5] and are beyond the scope of this chapter. It might be noted, however, that the most successful courses are related directly to the needs of the particular organization and to the specific problems of its staff. Lectures, therefore, must usually deal swiftly with the generalities of the subject and then get down to a discussion of problems and their solution, often via open-forum discussions in small groups. Some courses include outside writing assignments, followed by individual conferences. Others use individual conferences to discuss personal problems that have been revealed through a study of earlier reports.

The internal technical editor is in a particularly favorable position to conduct courses in technical editing if he has gained the respect of senior staff members as a trusted colleague. He is also in a good position to present a "first course" if he does this at the beginning of a new, concerted effort; backed by management; accompanied by his preparation of a good style manual; and to be followed by continued personal assistance to the writers who attend.

There are occasions and organizations, however, where the technical editor is wise to call in outside lecturers and consultants of known ability, since "a prophet is (sometimes) without (sufficient) honor in his own generation," and the voice of an outside expert might therefore be accorded more attention and respect, plus less chance of subsequent resentment of criticism. There are other occasions when respected technical supervisors of good writing ability make the best lecturers and conference leaders. Whatever the optimum combination, the good technical editor recognizes the need for internal courses on technical writing and usually plays an active role in their organization, conduct, and follow-up.

Day-to-Day Relationships

The internal technical editor can get off to a flying start by preparing a good style manual and conducting a well-prepared course on technical writing; but his most effective educational tool is the proper conduct of his day-to-day relations with the technical writers whose work he edits. These range from answering pop-in questions on the hyphenation of a complex word to carefully planned discussions of specific documents being returned to the authors after editing.

As mentioned earlier, the technical editor must always explain his editing—why he asks for changes or has made them. His reasons will usually be specific to the points in question, but—as he gets to know his writers better—they may also be related to the general writing problems of the scientists.

There is no better way to drive home a point than by illustration, so a sequence of edited documents usually provides the editor-author team with opportunities for a full-course treatment. Naturally, the editor must not lecture—must make his points as discussions with interested colleagues. He must suit these discussions to the writers' schedules, and must forego occasions when the writers are really pressed for time, but he must make his own time available. Hard as it often seems to find time for this type of educational effort, the technical editor who does not take the time to teach—who merely puts patches on a sequence of documents from a struggling author—will be doing a real disservice to himself, the writer, and their organization.

There are many other ways in which the technical editor can work constructively with writers on a day-to-day basis. He can circulate copies of pertinent, well phrased articles and pamphlets on writing that have been prepared by other organizations—documents which serve to reemphasize points which he has been making. He can invite editorial comments from writers on his own papers or those of colleagues, and thus improve rapport while imparting a better understanding of both writing and editing. He can urge his company to send certain writers to nearby short courses on technical writing, especially when no course of his own is imminent. He can cooperate with his organization's library in developing its reference collection on grammar, writing, and graphic arts, and can call pertinent books to the attention of writers.[1, 8]

The day-to-day relationships between editor and author thus afford a wealth of opportunities for helping technical men to improve their writing efforts. The efficient editor seizes as many of these as possible.

PROFESSIONAL ASPECTS

At this point it has been fairly well demonstrated that technical editing involves complex human relationships, including opportunities for educational efforts, in addition to the taken-for-granted ability to improve material submitted for editing. It now seems timely to consider some of the problems of the technical editor that relate to his status as a professional.

Organizational Status

Inside the organization which employs him, the technical editor may be variously classed as a representative of management, as an advisory-staff member, as a member of a service division, or as member of a project team. Each such status has its distinct aspects.

As a representative of management, the technical editor is endowed with a measure of authority in his work—an authority which he calls into play as seldom as possible. He is authorized to enforce management requirements on format and style, and often has definite responsibilities as to scheduling, budgets for related work, and document duplication (he often directs graphic-arts groups). He may have related supervisory responsibilities, direct or delegated. All this is true whether he is the editor of a journal, works in a book-publishing house, or is an internal editor for a governmental or industrial organization. Much of it pertains to the technical-project supervisor whose work also embraces extensive editorial control over the writings of his staff.

As a staff member, the technical editor represents management more indirectly. His authority is more advisory than direct, and he must usually persuade rather than command. He may still direct graphic-arts groups, and he often serves as a focal point for scheduling, chiefly as regards notices and records. He is still in a strategic position, however, since his is the role of a senior fellow worker. His accomplishments are directly related to his ability and the climate within the organization for which he works—its awareness of the need for good technical writing, and its willingness to take the necessary steps to aid technical employees with their writing.

Some organizations have placed technical editors in service divisions, likening their functions to those of safety engineers, personnel men, accounting departments, and others who work in auxiliary roles. Since most service divisions are largely nontechnical in function and personnel, the technical editors in such divisions sometimes seem to have a lower status in the eyes of technical-project personnel than do management or staff editors. This may reflect on their acceptance as co-workers, and may even limit their role to that of "clean-up men," editors who can be called upon for help or not as the author desires. Nevertheless, the dedicated editor can often surmount these difficulties and can carve his own niche in the esteem of the technical staff.

Some divisions of an organization—or even project teams—may be so large that they can efficiently use a technical editor as a member of the team, to schedule and expedite technical writing by other project personnel, to edit their documents, to handle illustration and duplication, etc. Such project editors often do a certain amount of writing of composite progress reports and the like. Their status is little different from their fellow project workers, and they are usually expected to have about the same backgrounds and to be able (when needed) to take part in other project work. They are in a good position to do effective editing if they become aware of the mores of their work.

Journal and book editors have in some respects a privileged standing—they are in a position to make or request editorial changes with the (usually unmentioned but real) implication that the document otherwise will not be published. They obviously exercise their power with discretion, since abusing it would work to their own harm and that of their organizations, but use it they do and must. Within their own organizations they have standings which vary with their experience, abilities, and responsibilities, but which are unique in that they are links in an editorial chain instead of *the* editorial link in a technical chain. Further discussion is not warranted here.

Professional Status

No detailed discussion seems warranted, also, on the right to call technical editing a profession. A recent paper by F. J. Van Antwerpen,[3] himself a former technical editor, points out that the word "profession" has seemingly "grown to include every activity of man for which special training and talent is needed," but that "engineers subscribe to the following criteria for judging a profession:

"1. It must satisfy an indispensable and beneficial social need.

"2. It must require exercise of discretion and judgment.

"3. It must be conducted on a high intellectual plane.

"4. It must entail group consciousness.

"5. It should have legal status."

Van Antwerpen quotes Vannevar Bush as stating that "the hallmark of a profession . . . is that its members minister to the people," and winds up with his own definition: "A profession is a group of men and women who develop conscious standards of work to which they hold themselves and one another responsible." By most (but not all) of these standards, technical editors (grouped perhaps with full-time technical writers) are members of a technical profession—or at least of a group which is on its way to becoming one.

Educational Requirements

The ideal technical editor, of course, is a technical man trained and experienced in the sciences and techniques of the field(s) in which he edits documents; educated and experienced in the use of words to transmit messages effectively; able and willing to sublimate his personal creative urge in the broader task of helping others to express themselves better; sincerely interested in the work of others, and in working with people; and able, on occasion, to represent management to scientists and scientists to management. Since few technical editors actually measure up to this exalted standard, it is not surprising that there is still some debate as to what compromises are acceptable.

Most authorities now agree that the technical editor ought

to be a graduate engineer or scientist—that the English or journalism major without technical training is under too great a handicap when it comes to that type of editing which often calls for a form of technical collaboration rather than grammatical correction; i.e., *technical* editing. Obviously, however, the good technical editor must also be able to write and to edit properly—to understand and explain syntax and semantics, skills which usually require extensive education in grammar and composition either before or after he begins editorial work.

Few technical editors have had formal training in psychology or in management practices, so most of the other ideal requirements must come naturally to the good editor or must be acquired through experience and hard study. Short courses in technical writing designed for professionals are consciously or unconsciously helping neophytes in many of these areas. Some of the growing number of degree courses in technical journalism take them partially into account, but it is too early to judge with what effectiveness.

Most of today's good technical editors did not set out specifically to enter their field. Instead, they first became engineers or scientists, demonstrated their writing or editing abilities, became interested in and dedicated to technical editing, and thus grew into their work like Topsy. However, this process is quite apparently producing too few technical editors to meet modern needs, hence the current efforts to college-produce technical journalists. It would appear that *both* routes will continue to be followed.

The training of internal technical editors is a subject unto itself. It is described in Chapter Seven.

OTHER ASPECTS

All this emphasis on psychological, educational, and professional aspects of technical editing may have made the technical editor seem too distinct a person—too removed from the ranks of laboratory and operational scientists and even from those of technical writers. In the present state of the profession, he certainly is not. Many technical editors still do their editing along with service as laboratory scientists, technical writers, and special librarians, especially in small organizations. Even in large organizations the internal technical editor is sometimes called upon to project his knowledge of management and science—and writing, naturally—through the media of ghost-written documents as well as those under his own by-line.

But this much *is* clear—the scientist, as such, is not necessarily a good technical writer, and the good technical writer may not have the interests and abilities to be a good technical editor. Hopefully, this chapter, if it has done anything, has established that there are unique psychological, educational, and professional aspects to technical editing, and that these merit serious consideration by all concerned.

REFERENCES

1. Davis, D. S., "The Educational Role of the Technical Editor," *J. Chem. Educ.*, **33,** No. 2, 57-59 (1956).

2. Rathbone, R. R., "Engineer-Writers Are Made, Not Born," *Res. and Eng.*, **3,** No. 1, 22-23, 43 (1957).

3. Van Antwerpen, F. J., "Professional Societies and Professional Standards," *Special Libraries,* **48,** 251-54 (1957).

4. Waldo, W. H., "Instructing the Report Author," Paper presented before the Division of Chemical Literature, 125th Meeting Am. Chem. Soc., Kansas City, Missouri, 1954.

5. Waldo, W. H., "Teaching Report Writing to Professional Chemists and Chemical Engineers," *J. Chem. Educ.*, **33,** No. 2, 59-61 (1956).

6. Weil, B. H., "Style Manuals," Chapter Nine in *The Technical Report, Its Preparation, Processing, and Use in Industry and Government* (B. H. Weil, Ed.), New York, Reinhold Publishing Corporation, 1954.

7. Weil, B. H., and Lane, J. C., "Some Psychological Aspects of Technical Reports," *Chem. Eng. News,* **34,** 6244-46, -48 (1956).

8. Weil, B. H., and Poland, L. B., "Library Aids to Technical Writing," *Special Libraries,* **49,** 11-16 (1958).

SELECTED BIBLIOGRAPHY

9. Cortelyou, Ethaline, "Some Qualifications of a Technical Editor," Paper presented before the Am. Documentation Institute, Cleveland, 1954.

10. Miller, Robert C., "The Care and Training of Authors," *Science,* **121,** 526-28 (1955).

11. Samstag, Nicholas, "Copy Editing: A Personal Equation," *Printers' Ink,* **242,** 43-45 (Jan. 16, 1953).

12. Southgate, M. Therese, "The Editor's Duty: Turning Out Good Authors. Outline for a Report Writing Course," Paper presented before the Am. Documentation Institute, Cleveland, 1954.

13. Thistle, Melville W., "Barriers and Approaches to Scientific Writing," Paper presented before the Assoc. Tech. Writers and Editors, New York, 1956.

14. Wall, Florence E., "Requirements and Responsibilities of a Technical Editor," *J. Chem. Educ.,* **30,** 516-21 (1953).

Part I

EDITING INTERNAL DOCUMENTS

Chapter Two

Editing Technical Reports

IRLENE ROEMER STEPHENS
Research Librarian, Celanese Corporation of America
Summit, New Jersey

and

P. M. REYLING
Supervisor, Technical Publications Department
Oak Ridge National Laboratory
Oak Ridge, Tennessee

Over the last few decades, most research and industrial concerns have placed more and more emphasis on the preparation of technical reports to be read by internal audiences and, on occasion, by outside individuals.

For many years it was considered sufficient merely to document various findings for the record, with little attention to style, format, or sometimes even to technical accuracy. However, as greater effort was directed toward research, and as management became more conscious of the value of good public relations (internally for the development of individuals and the proper evaluation of projects, and externally as evidence of technical proficiency), special attention was given to the preparation of better reports and, inevitably, to their editing. With the flood of technical reports on every conceivable subject, moreover, readers with any choice began to select for detailed review only those reports which were carefully prepared.

The purpose of a technical report is not only to record the results of an investigation, but also to present the details and reasonings associated with obtaining them in a form which is easily referred to and readily understood by the audience for which the report is intended. From this definition it is apparent that there are a number of different kinds of technical reports. For example, the report of the bench worker is different in format and general content from the report of the research director to management. However, whether we are considering the report of the bench worker to his section head, the report of the section head to his division head, the report of the division head to his director, or the report of the director to the vice president in charge of the division, the role played by the technical-report editor is much the same.

The remarks which follow are made in the same frame of reference as those in the related portions of Chapter One. Here they bear directly on reports.

AUTHOR-EDITOR RELATIONS

To facilitate the production of accurate and well-written technical reports, many research and industrial groups have set up editorial units. It is the function of the editors in such units to help authors turn out reports which are readily understandable as well as technically accurate. It is their responsibility to see that ideas are clearly and simply expressed, to avoid ambiguity, obscurity, and error.

The competent report editor must do his editing primarily

in terms of the readers of the report and only secondarily in terms of the author, yet it is with the author that he must first establish a satisfactory relationship. The editor's primary interest is to ensure that the author's intended meaning is conveyed accurately, efficiently, and effectively to all readers for whom the report is being written. In effecting this, however, he must be aware that the author has put his best efforts into preparing his manuscript—that the author's typescript may represent months or years of meticulous research, preceded by long hours of literature searching and followed by a painstaking effort to present his findings properly in written form. The author is therefore understandably proud of his finished manuscript, or at least feels that it is good enough for the purpose for which it was prepared. He may therefore be very sensitive or unreceptive to criticism. While he knows that the editor intends to be helpful, he may feel that, under the guise of helpfulness, the editor is being authoritarian and insistent on changes of questionable value.

Most assuredly, it is the business of the editor to maintain standards. How far, then, should the editor compromise with the author, and how far should he insist on those alterations which he believes desirable for the improvement of the document or necessary for the maintenance of standards?

Many of the difficulties which arise between author and editor can be avoided, and the work of each made easier, if the editor will tell the author what he wants in advance. It is for this reason that many—indeed most—editorial groups have found it desirable to create style manuals. These manuals not only set forth the standard format desired, and rules of good English, but they also reflect the desires of management. They must themselves be prepared in accordance with very high standards and must be readily understandable to all for whom they are intended.

The possibility of author resentment of editorial changes

is minimized if the editor approaches the author as an assistant or colleague, sincerely interested in turning out a final product of which both will be proud. If the editor has questions about possible changes, he might well ask: "Could you help me with some things I don't understand?" For best relations the editor does not pose as an expert in the technology involved, even if he may be well versed in the subject field. The report is a detailed account of the work of the researcher. No one knows that research better than the individuals actually involved. Therefore, the author must usually have the final word as to the presentation of details.

Using the vital cushions of tact and diplomacy, the editor must establish himself as an expert in assisting the author in turning out technical information in understandable form. The experienced editor will achieve a high degree of success and cooperation by submitting suggestions for improvement, by selling rather than by telling.

It is sometimes possible to close the gap between the editor and the author by setting up an arrangement in which authors are assigned, for specific periods of time, the responsibilities of editors, at least to edit work for technical accuracy. Although it is then not the technical man's responsibility to ferret out mistakes in format, grammar, punctuation, etc., he usually begins to note these errors. The technical man who has cooperated in this kind of technical editing almost always pays more attention to details of standardization when he writes his next report.

This review of reports by technical people other than the author can be initiated and carried out in several ways. If the plan is agreed to, in principle, by both the research and editorial groups, all researchers can be asked to cooperate. It can be pointed out that the technical man who is editing for technical accuracy benefits by being able to read the report earlier than usual. The very nature of his editing assignment, moreover, means that he must pay close attention to technical details, which may contribute to more effective use of the report in his own work.

Normally, it is advisable to have the identity of the technical reviewer remain unknown to the author. This arrangement is similar to that employed by scientific periodicals.

Periodic meetings of report editors and technical personnel provide occasions for joint discussion. Often, as a result of these discussions, the style manual may be modified or clarified, and additions may be made. In semiformal discussion groups such as these, much can be accomplished in improving both reports and the relation between authors and editors.

THE STYLE MANUAL AND ITS USE

The use of a style manual for reports avoids a great deal of discussion and needless changes. It puts the editor in a position that can be defended by logic.

When an organization first prepares a style manual, or when it feels that a drastic revision is needed, the occasion may be used not only to establish a standardized approach to report writing, but also to win the cooperation of the individuals in the research or operating groups. Obviously, the manual must first reflect the policies of management. Beyond that, however, the editor can solicit suggestions and can use them as a basis for the manual. Furthermore, through a statement in the preface of the manual and by periodic follow-ups (both personal and formal), the editor can remind the report author that the manual is a tool intended to help both the author and the editor. Under certain circumstances (but not in large organizations), the editor can even ask for suggestions as to changes and additions. Even if the author never finds occasion to make such suggestions, he knows that the business of making the manual as complete and effective as possible is a joint responsibility. Under any of these circumstances, the editor can refer the author to a manual which the editorial group and the research or operating groups have jointly prepared, not to a manual set up by the editors to enforce rules agreed on by editors and management alone.

A complete style manual will cover format, grammar, abbreviations, typography, graphic arts (including photography), a glossary of unusual technical terms, and any special requirements. It is an important tool in standardization. However, the editor, to apply it effectively, must have a ready knowledge of grammar, spelling, punctuation, capitalization, foreign words and phrases, international nomenclature rules, and special vocabularies of various sciences. He must accept the fact that most words have several meanings or exact shades of meaning. He must learn to identify these meanings with precision and then sort them into well-defined compartments by subject specialties. He must keep abreast of the times and be aware that language is a living thing and that current usage changes.

The editor must assume that the author is technically competent and, furthermore, that the author recognizes the importance of effective graphic communication. If the author is not following the prescribed format and rules of presentation as included in the style manual, the editor can very easily refer the author to this manual for these mechanical errors. Matters of literary style, however, are considerably more delicate. It is not the function of the editor to rephrase the report in his own literary style. Usually, this style is not the important aspect of the report. The report must be understandable. It must be readable. It should be written in clear, concise language.

The foremost consideration is technical accuracy. Standardization of format and grammar then aids in making the report more understandable. Certainly the report must be presented in such a way that the technical details are easily interpreted by the audience for which the report is intended.

THE ROLE OF THE EDITORIAL SUPERVISOR

In small-scale operations the editor is generally the supervisor and may be responsible for many widely different functions. However, where large-scale operations are involved, the editorial supervisor plays a key role in producing quality reports.

An editorial supervisor must obviously have a strong background in the detailed work which editors normally perform. In addition, he must have an extensive knowledge of the arts required to complete finished reports, namely, the graphic arts and reproduction. A working acquaintance with psychology and diplomacy is also a necessity. Judgment, as related to the quality and appearance of finished reports, is an important requisite.

The editorial supervisor in such a position is not usually responsible for the technical accuracy of reports, because either the author or his line supervisor should make the final decision on content. However, definite responsibility exists for a cooperative effort to obtain maximum readability by tactful handling of author-editor relations.

Another important function concerns the scheduling of reports. Processing time must be held to a minimum in order that the finished document may be distributed as soon after development of the data as possible. For example, a quarterly progress report that is distributed six months after the close of the reporting period loses much of its value. Close coordination of the efforts of editors and of graphic-arts and reproduction personnel will result in a timely, finished product.

LARGE-SCALE EDITORIAL ORGANIZATION

For most effective operation, the editorial unit must be well organized. Long-term expansion needs as well as immediate needs of the unit must be considered in setting up such groups. Consolidation or close coordination with other communication groups is usually highly effective, whether these groups are large or small. This can be judged from the organization chart shown in Figure 2.1, representing a close-knit group designed to service a research laboratory of approximately 4,000 people (including both scientific and supporting personnel). The salient features of this setup are: (1) a single division head, placed on the same level as other division heads within the organization, (2) logical separation of functions by sections, and (3) availability of a group of sufficient size to be capable of top-quality work on a continuous basis.

For smaller laboratories or plants, fewer people are needed for each function, but it is desirable to retain a single head for the various groups which perform the functions shown in Figure 2.1. In small laboratories or plants, moreover, a very desirable approach to ensuring high-quality reports and printed matter is to have the over-all group, whether division or department, report directly to the laboratory director or plant manager. With this arrangement the publishing group in general, and the editorial group in particular, can reflect more accurately and with greater efficiency the views of the director or manager. The administrative position illustrated in Figure 2.1 is on a par with that of other division heads. Therefore, internal differences and delays are minimized.

Figure 2.1 shows the various sections directly involved in publishing, as well as those sections that are closely allied with such work. Editors then have working relations not

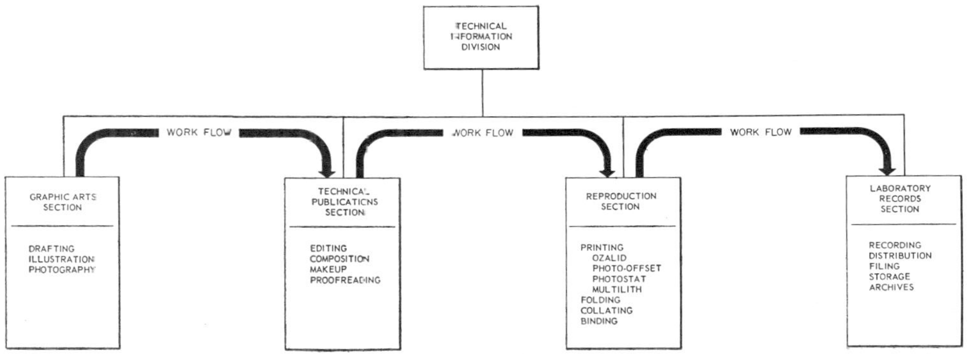

Figure 2.1. Technical-information-division organization.

only with authors but also with all the sections shown. The processing of actual reports is shown by the heavy arrow labeled "work flow." The Graphic Arts Section and Publications Section work jointly in producing finished copy and illustrations. The master copy which results from the joint activities of these two groups requires only printing when given to the Reproduction Section. The editor is consulted relative to the details of reproduction. In these relations, as in those with authors, tact and diplomacy are desirable on the part of the editor, as mentioned in Chapter One.

The operations of the Technical Publications Section will vary in industry and research. In general, however, in an organization such as the one illustrated, where various sections are utilized in processing reports, the procedure in editing reports is fairly standard. The editor first receives a draft of the report from the author, in which the author has presented the technical details completely. On receipt of this draft the editor checks the illustrations and sends them to the Graphic Arts Section with a desired completion date. The time period required for completion of final illustrations then allows time for editing of the text of the report, reviewing this with the author, "composition" (typing, Vari-Typing, etc.), and proofreading.

As the editor progresses in his handling of the report, he checks tabular material and illustrative copy against the body of the report, reviews references for accurate citation, and, if the report is to be Vari-Typed or printed, marks the copy for type face and size. The resulting edited report is then ready for review with the author. After the author and editor have reviewed the report, and final agreement has been reached, the report is ready for composition. The editor then follows the progress of the report through composition, proofreading, and makeup.

SMALL-SCALE EDITORIAL ORGANIZATION

Editing of reports in a smaller organization than the one used in the illustration is similar in its over-all function but much less formal in practice. In a small organization there will probably be only one editor. It is possible, too, that the responsibilities of the individual may extend beyond the editorial functions which such a title implies. The editor may actually be an administrative assistant to a research director or to a department head. He may write reports, edit reports, and write speeches and articles for research people and management groups. He may index reports and disseminate them. As a matter of fact, he may also be the organization's literature searcher, patent agent, or special librarian. The actual title of the "report editor" may thus be almost any title applied to individuals with responsibilities in literature work, one in which editorial responsibilities are not necessarily implied.

Whatever the title and other functions, however, the implications of editorial work are the same. Diversified activities need not detract from the effectiveness of the editor's role as long as time is provided for fulfilling the editorial responsibilities. As a matter of fact, contacts between authors and editors in connection with other activities may tend to make editorial conferences more informal.

SUMMARY

In summary, then, it is clear that when editors are assigned the responsibility of assisting authors, reports can be issued which reflect more accurately the views of management relative to the purpose of reports and the role which these reports play in the research organization. When an understanding is reached between the editor and authors as to the importance and advantages of teamwork, technical reports reflecting creditably on both author and editor are produced. More important, the results of such cooperation are reports that are effective. They are technically accurate, grammatically correct, reflect the views of management, and are understandable to the audience for which they are intended. As a matter of fact, they are readable.

Excellent research poorly reported may sometimes be interpreted as of less worth than less significant research that has been well reported. It is easily apparent, then, that the careful editing of reports adds to the subsequent utilization of the results reported.

Chapter Three

Internal Editing of Technical Papers and Articles

BURR J. FRENCH
Technical Analyst-Writer, Information Services Group
Ethyl Corporation Research Laboratories
Detroit, Michigan

Quality control of products has long been important to manufacturing companies. Automobile manufacturers, for example, realize that the malfunctioning of a part will be a source of owner discontent, and may result in his turning to another company when shopping for his next car. As insurance against such occurrences, manufacturers have established quality-control groups whose responsibility is to provide a continuous check on the quality of the various products made by their companies.

More recently, many organizations have recognized that quality control is also essential in the production of technical papers and articles.[1, 4, 5] If the article is well written and informative, the company will profit from the presence of its name under the author's byline. On the other hand, a poorly written, muddled article will be rejected by most journal editors or, if published, may induce a negative reaction in the average reader. This chapter deals with some of the means of achieving adequate internal quality control of technical papers and articles.

Ideally, the responsibility for quality control is vested in a technical editor and in a policy-review committee which includes top-management personnel. Since the review committee usually performs the initial job of screening, its function will be considered first.

THE REVIEW COMMITTEE

The functions of a review committee can perhaps be best illustrated by a specific example. At Ethyl Corporation's Research Laboratories, the Technical Publications Committee is composed of management representatives from the research and engineering, manufacturing, advertising, sales, public relations, and patent divisions. This committee is empowered to approve plans to present or publish papers on specific subjects—first submitted in outline form or as abstracts—and then to review the final versions prior to release.

In their review, committee members apply three primary yardsticks to each paper or article:

(1) Does it constitute a useful addition to knowledge in its field?
(2) Does it contain confidential information or information that might endanger present or future patent applications?
(3) Does it uphold the company's reputation?

The first two are the controlling factors; the third is a desirable byproduct.

In large organizations which have allied activities at

several locations, technical-staff members are often asked to review papers at the same time as management; i.e., the personnel of a review committee varies with the subject of the paper. All members of a review committee are free to comment on the manner of presentation as well as the technical and policy implications of a given paper—and they often do. By itself, however, this has rarely provided adequate quality control.

Incidentally, this discussion assumes that a paper has been reviewed and approved for technical accuracy and interpretation by the author's immediate supervisor(s), and perhaps colleagues, before it is submitted for quality control and policy approval. In small organizations this review may be all the help an author can get.

THE EDITOR'S ROLE

After a paper has passed the tests of supervisor approval and policy survey, it is ready for quality control by the internal technical editor. This naturally includes the checking of spelling, punctuation, and sentence structure. More importantly, however, the internal editor of a technical paper or article must continually ask himself the following questions:

(1) Is this paper or article aimed at and written for the prospective audience?
(2) Will this paper or article secure and maintain listener or reader interest?
(3) Does this article fulfill the requirements of the society-sponsored or industrial journal in which it is to be published?

Incidentally, these three questions point up the basic differences between technical papers and articles and the technical report. In general, the writer or editor of a technical report does not have to be concerned with sizable variations in audience, for a report is normally designed to supply specific and complete information to people who are in the same line of work. By comparison, a technical paper or article is more often than not destined for a more general audience, and the needs and interests of that audience must be anticipated. Also, whereas a technical report is often "required" reading, a technical article is competing for the reader's time with other articles in the magazine.

In other words, the report writer has a captive audience, while the writer of a technical paper or article must capture and hold his audience. This does not imply that quality control of technical reports is unnecessary, for the message may escape unless the report is well-organized and clearly written. The point is that writing or editing with the reader in mind is even more essential when preparing technical papers and articles.

Prospective Audience

Constant awareness of the prospective audience is vital in the editing of a technical paper. For example, consider a research program in which special instruments and techniques were used to study the factors affecting spark-plug fouling. Out of this program could come several different papers, each aimed at a specific audience. The results, coupled with a brief description of the instruments and techniques, could be documented in a paper for the Society of Automotive Engineers. On the other hand, a paper for the Instrument Society of America would deal largely with the design and construction of the instruments employed. Or a paper delivered to the American Chemical Society could comprise a discussion of the compounds in the deposits formed on the spark plugs, and their solid-state reactions.

Not only the needs and interest of the audience, but also

its technical background must be considered. For example, chemistry and mechanical engineering are closely intertwined in the study of engine-fuel relationships. Thus, some mechanical-engineering terminology will inevitably creep into a paper on, say, the mechanism of the antiknock action of tetraethyllead. Care must be taken to define this terminology or to establish its meaning by the context in which it is used.

One of the greatest pitfalls in technical writing is the author's comprehensive knowledge of his area of research or engineering. Familiarity with his work often leads him to assume unconsciously that the potential reader has been "standing at his side" while the research was being conducted, and is therefore fully aware of background facts and theories. The author may thereupon neglect to include all the material necessary to orient the reader and to present a complete picture.

The internal editor of a paper is in a good position to review a paper for correctness of approach. Unless he is unusually close to the work reported, he is himself part of the author's audience, and he should be skilled in projecting himself into the shoes of other potential readers. That is why the editor usually reads a draft clear through, for over-all impact, before he gets down to a close review of each word, sentence, paragraph, and section.

Reader Interest

Just aiming a paper or article at an audience is not enough; the editor and author must work together in an attempt to ensure a favorable reception. Here, organization of the paper is the key. Unlike technical reports, which are usually written to conform to quite rigid patterns, the technical paper or article is a rather flexible animal. However, there are certain basic principles which apply to all technical articles and to most technical papers.

Perhaps the most important part of a technical article is the first paragraph, or lead. It is here that the author must capture the reader.* If he fails to do so, the rest of the article may be useless. A sound approach that can be used for many articles is to summarize the article in two or three carefully chosen sentences. In effect, the lead tells the reader what the article will do for him. One word of caution is necessary, however; the lead should not promise more than the article will provide.

Once the lead has been established, the rest of the article serves as amplification. With reader interest whetted, the next step is to get the story across. A successful approach, one that can be applied to most articles, is to discuss the most important point first, followed by others of decreasing importance. Thus, the reader will have received the meat of the story if he is dragged away before the end.

This method of approach—somewhat akin to technical press releases—may have to be modified to suit the policies of specific journals and societies. Some of these require a more formal approach—statement of the problem; description of the techniques and equipment used in its solution; reporting of the salient data obtained; then—and only then—discussion of its significance. Even here, the author and the editor can stimulate interest by use of a good lead—by a clear description of the importance of the project and the field in which it falls—and can hold this interest by proper progression throughout the paper.

Maintaining interest obviously implies the application of writing techniques that carry the reader smoothly from one sentence to the next and from one thought to another. The editor must therefore look for and correct involved sentences, wordiness, and jargon that serve as roadblocks to

---

* As editors of magazines point out, the title or heading of a magazine article should be carefully selected to attract the reader. Nevertheless, the lead must complete the job of capture.

the reader's comprehension. He can also help himself on future assignments by introducing his authors to writing aids such as the works of Robert Gunning.[2]

Matching Journal Specifications

Most society-sponsored and industrial journals have established definite editorial rules. These are published by them from time to time, either separately or in issues of the journals themselves, and can usually be obtained on request. Indeed, the company editor can provide a useful tool for both prospective authors and himself if he will compile a reference binder containing the editorial-policy rules of the various journals that service his company's field of interest.[5]

For the most part, such rules emanate from the basic aim of the technical journal—service to its readers. Rules dealing with nomenclature, literature citations, and the preparation of tables and illustrations are laid down as aids to maintaining the professional qualities of the journal. In their desire to serve the reader, many journal editors place a maximum limit on the length of an article— 4,000 words is a common limit. Even where no limit is indicated, editors tend to favor the shorter of two articles, assuming that other factors such as potential readership and coverage of the subject are equal.

There are some good reasons for this. It is well known that readers in general resist long articles. Out of this resistance such magazines as *The Reader's Digest* have grown and flourished. Also, a magazine which has a low ratio of editorial content to advertising pages cannot afford to expend eight pages on one article where two four-page articles on diverse subjects will probably reach more readers. It is true that some feature articles may consume 10 to 12 pages, but these are the exceptions rather than the rule.

Almost every article can be trimmed and boiled down without loss of vital information. Unfortunately, many engineers and chemists tend to use six words where one will do. Judicious pruning of these useless words both shortens an article and improves its readability.

Drastically boiling down an article is another matter—one that usually requires delicate handling of the author. Just as a man's legs should be only long enough to reach the ground, a technical article should be only long enough to reach the reader. However, editing for length requires a fine sense of balance. Like the tightrope artist, it's bad to fall either way. On one side is excessive detail and "writing down to the reader"; on the other side is the omission of facts required for a complete understanding of the article.

Here is where the lead paragraph becomes useful to the editor. As previously noted, this lead has promised the reader that he will receive certain information. If a sentence or a paragraph conveys thoughts that are not essential to this thesis, the editor may ruthlessly cut it out. Such nonessential matter not only increases the length of an article, but may also create a stumbling block for the reader by diverting him from its thesis.

THE EDITOR AND THE AUTHOR

Preceding portions of this chapter have dealt with some factors that the internal editor must consider when judging a technical paper. Of almost equal importance is how the editor actually works with an author to achieve a sound technical paper or article.

In the editor-author relationship, it must be remembered that the whole philosophy of editing is service, even when the editing is mandatory. Consistent with this philosophy, the editor should not impose his style of writing on the author. At the same time, the editor must not hesitate to suggest alternate wordings where the author's statements are not clear or are actually misleading. If necessary, the editor must also propose changes in organization that will strengthen the paper or article.

Here is where the editor must don the coat of a diplomat, for many authors, especially inexperienced ones, are sensitive about every paragraph, sentence, and word. Fortunately, there are others who welcome constructive criticism; they are like rays of sunshine in an editor's life. However, it is not within the scope of this chapter to discuss the psychological aspects of the editor-author relationship (these are dealt with in Chapter One), except to say that common sense seasoned with a lot of patience helps a great deal.

One step that would lighten the editing job would be an informal session with the author before his first word is put on paper. With inexperienced authors, the editor could then explain how technical papers or articles differ from the technical reports they are used to writing, could discuss the identity of intended readers, could place emphasis on the need for conciseness and clarity, and could even go into the possible organization of the paper or article.

Such a session may not be possible. In this event, a review session with the author after the article has been edited will pay dividends in the future. Editorial changes must be checked with the author anyway, so the editor will do well to spend a little extra time in giving the author a solid background on why the changes were made.

There are two precepts of writing that every editor would do well to drum into authors at every chance. They are:

(1) Think before you write.
(2) Do not write so that you may be understood; write so that you cannot be misunderstood.

The first precept is one that Norman G. Shidle has been driving home for many years.[3] His premise is that clear thinking must precede clear writing; that one cannot write clearly unless his thoughts are well sorted and organized. Although there can hardly be any disagreement with this, many authors start to write before they are ready. This almost inevitably results in an excessive expenditure of hours and in a bad piece of writing.

The second precept may sound a little like "gobbledegook." And yet lack of its observance is a frequent pitfall for the scientist or engineer who is an expert in his field. Complete familiarity with a subject may lead to a state of forgetfulness in which the writer fails to recognize that a phrase or word may mean different things to different people, or nothing to some people.

* * *

This chapter has attempted to summarize some of the facets of internal editing of technical papers and articles. One fact stands out. The technical paper or article must sell itself to the listener or reader. It is the editor's responsibility to assist the author in achieving this objective.

REFERENCES

1. Davis, D. S., "How to Use an Editor," *Chem. Eng.*, **56,** No. 12, 101-03 (1949).

2. Gunning, Robert, "The Technique of Clear Writing," New York, McGraw-Hill Book Company, Inc., 1952.

3. Shidle, Norman G., "Clear Writing for Easy Reading," New York, McGraw-Hill Book Company, Inc., 1951.

4. Weil, B. H., "The Quality Control of Scientific and Technical Papers and Reports," Paper presented before the Division of Chemical Literature, 118th Meeting Am. Chem. Soc., Chicago, 1950.

5. Weil, B. H., "What Organizations Are Doing to Improve Papers," *Chem. Literature,* **7,** No. 2, 3 (1955).

Chapter Four

Editing Organization-Sponsored Magazines and Newsletters

JOHN C. LANE
Superintendent of Information Services
Ethyl Corporation Research Laboratories
Detroit, Michigan

Lack of a clearly defined purpose makes many sponsored publications (technical and nontechnical) much less effective than they could be. This chapter will show that many of the problems involved in editing such a publication can be resolved if its purpose is clearly spelled out.

In some publications, for example, there is the question of the right balance in coverage between company activities and employee activities. Certain magazines give you an almost perfect count of baby production by employees' wives but practically no information on corporate health and output. In others, the only names ever mentioned are those of the chairman, the president, and the treasurer.

Now, either of these two extremes could possibly be justified. That would depend on whether the publication was intended exclusively for employees or solely for outside circulation. However, a searching study of over-all purpose would probably indicate a lack of desirable coverage in each case. Employees want to know something about company plans and progress—after all, this is bread-and-butter news. On the other hand, significant news about employees' accomplishments will make outsiders feel that the company is human and therefore probably put more faith in its products.

Unfortunately, about 60 per cent of all company-sponsored publications do not have a written statement of their aims and objectives. At least this was the finding in a 1956 survey answered by 94 industrial editors.[1]

BASIC PURPOSES

Probably no two sponsored publications have exactly the same over-all purpose. If they did, they would have to have identical aims and be pointed at exactly the same proportions of several different types of readers.

However, all publications have this general purpose in common: to educate, inform, and entertain their readers. The first two functions, educating and informing, consist of telling the reader things he ought to know plus things he presumably wants to know. The third function, entertaining him, is necessary to the extent required first to attract his interest and then to hold it.[2]

The specific objectives of publications differ, depending on whether they are intended for employees or for outside distribution. The following seven points have been listed as objectives for *employee publications*:[2]

(1) To humanize the company to the employees.
(2) To integrate management and employee interests.

(3) To integrate the staff with the operating personnel.
(4) To recognize employees' accomplishments.
(5) To provide information on the company, its products, and its place in the industry.
(6) To bridge the gap between home and job.
(7) To explain and interpret economic facts.

Seven suggested objectives for *publications going to customers* are:

(1) To build confidence in product or service.
(2) To build confidence in the company.
(3) To explain company policies and programs to customers.
(4) To suggest marketing ideas and practices.
(5) To stimulate sales spirit.
(6) To report field news.
(7) To discuss and interpret business trends.

Both of those lists cover fairly general objectives which might fit a large number of publications in each class, internal and external, technical or nontechnical. However, the customer-publication aims seem tailored to companies which produce a product for wholesale distribution, judging from the accent on marketing ideas, building sales spirit, etc.

A different set of objectives might be established for the publication of an organization doing sponsored research. Here the basic purpose would be to sell research services. Thus, the publication should be designed with the following objectives in mind:

(1) To sell the idea that sponsoring a research project can often be more economical than staffing and equipping to do it within the company.
(2) To build confidence in the research organization's personnel and facilities.

(3) Preferably to do this by factual stories of projects that paid off to their sponsors.

Even when there is agreement on the over-all purpose and the fairly detailed objectives of a publication, there still may be different opinions on how to pursue each objective. These opinions can apply to and determine any aspect of the publication, even a physical aspect such as typography.

For example, the director of a sponsored-research organization once asked his editor to design a publication which would use a low-key, factual, and convincing approach to help sell the organization's services. Figure 4.1 shows the typography which the editor chose for the top of the cover of this publication. As can be seen, the editor chose an old style of type. This was to make the readers feel, on first sight, that here is a long-established organization, and to imply that this is a research journal instead of just another promotional magazine. Naturally, the subject matter of the journal was chosen to confirm and drive home these impressions. That is, primary emphasis was placed on research being conducted, and on results obtained.

Later, a new director felt that a more forceful selling job should be done. This called for "streamlining" the journal. Figure 4.2 shows the typography selected for the top of the cover. A modern type and a color block were used to convey the desired impression. In keeping with the desired purpose, the text placed greater emphasis on the growing potentialities of the research organization. That is, more space was given to building programs, acquisition of new research facilities, and similar subjects.

Finally, under a third director, the cover typography was again changed, as shown in Figure 4.3. A modern but somewhat more conservative type was used on the cover. The subject matter of the publication varied roughly in proportion to the changes made in typography. In other words,

Figure 4.1. Masthead for a low-key, factual research bulletin.

Figure 4.2. Top of cover for a research bulletin intended to do a forceful selling job. A color block was employed.

Figure 4.3. Top of cover for a research bulletin, stressing modern approach but conservative type.

each editor was working toward the same general purpose, but was using a somewhat different approach.

EFFECT OF PURPOSE ON EDITING DECISIONS

We have already considered briefly how some editing decisions are determined by the purpose and objectives of publications. We have seen that these factors affect the desirable balance in coverage between company activities and employee activities. We have also seen examples of how different approaches to the same general purpose dictated choices of subject matter and typography.

Actually, almost every aspect of a publication is determined by its purpose, or at least should be. For example, the most effective ratio of feature articles to news items obviously depends on the publication's objectives. In many cases, information on company activities, the work of various departments, research and development programs, benefit plans, etc., can best be conveyed in detailed feature articles. Magazines whose purposes are best served by supplying such information may consist almost entirely of articles. Other magazines have as a main objective the reporting of field news to dealers and customers. These may have a high proportion of news items and correspondingly fewer feature articles.

Purpose also determines how often a magazine should be published. We have seen that purpose determines the ratio of news to feature articles. This ratio, in turn, determines publishing periodicity. To be current, a magazine devoted to news should be published weekly, or at least monthly. On the other hand, most feature articles are not so perishable and can be published in a monthly, bimonthly, or quarterly with about equal effectiveness. A magazine devoted 50-50 to news and features might well be published either weekly or monthly. In other words, the need to keep the news content current is controlling in magazines containing any substantial proportion of news.

A publication's balance between text and illustrations also depends largely on its over-all purpose and its detailed objectives. We have all heard the proverb: "A picture is worth a thousand words." However, as one writer has pointed out, this proverb is remembered only because it was put into words or text which could be translated into many languages.[2] As he says, "Text can be memorized and repeated by one person to another, whereas a picture can be described only sketchily and never with complete accuracy. The only way to think about picture-text balance is to remember that each serves a purpose, text being better to tell one kind of story and pictures another. . . .

"Caution points are these: If the percentage of text is high, then the *attraction value* of the publication may be lower. Conversely, if the percentage of pictures is high, the *remembrance value* of the contents may be lower." Here again is a case of considering purposes and then choosing the proper methods. As examples of picture-text balance chosen to suit a purpose, we might compare *Life* and *Time*. When a friend says, "Say, did you see in *Life*," he usually is referring to a picture. The same introduction mentioning *Time* probably precedes a discussion of a text item.

The intended reading audience and the purpose of a publication are so closely related that there is no way of saying which determines the other. Obviously, the purpose is to educate and inform a certain group of people. The results desired will determine the audience which should be reached. You would not send a journal aimed at selling chemical-research services to an audience of hardware retailers. Nor would you send a magazine of hardware field news to an audience of chemical-company research directors.

On the other hand, the intended audience determines how the purpose of a publication can best be accomplished, and may largely shape its objectives. This is obvious in the case of an employee publication, where consideration must be given to such factors as ratio of male to female readers, ratio of office to plant employees, the varying educational levels of readers, etc. Publications given wide outside-the-company circulation must also offer something to each reader. This can be done by balancing the issue in subject matter—analyses of trends for executives, technical articles for scientists and engineers, and various types of feature articles, field news, and the like for other readers. Wide appeal can also be given, even to technical and economic articles, by skillfully clear writing and by concentration on the practical, universally interesting aspects of the subject.

Mention might be made here that audience and purpose, together, should determine the balance between story length or brevity. The readers of *Fortune* magazine presumably are, for the most part, highly educated people who are willing to read long and detailed articles for the sake of getting the whole story. Also, *Fortune* is very well written and eminently readable, a fact which makes its high proportion of long articles more palatable to its audience.

Most publications which consist almost exclusively of long articles are likely to repel their readers. On the other hand, exclusive use of short articles also repels readers, since they hardly get interested before they are through—possibly with unanswered questions in their minds. A balanced mixture of long and short articles will attract and hold the most readers.[2]

FUNCTIONS OF THE EDITOR

So far, we have considered problems whose solutions depend largely on knowing the purpose and the audience of the publication. Management usually works with the editor in determining both the purpose and the audience. Thus, the editor is helped or directed in a number of decisions affecting the publication.

Now, let's consider some functions or responsibilities for which the editor usually is solely responsible.

(1) He plans each issue, balancing its subject matter so that the average reader will always find something of interest.
(2) He makes news-gathering assignments to his "roving reporters" or field editors, and follows up on these as needed to meet deadlines.
(3) He lines up appropriate authors for planned articles and/or follows up on leads which may uncover suitable articles.
(4) He makes certain that the desired subject coverage is understood by the author, and offers suggestions (sometimes as written guides) on desired form, length, slant, type of illustrations, and other factors.
(5) He schedules these articles well in advance, and checks each author's progress routinely on a pre-established schedule.
(6) He makes certain that each article submitted is constructive and that it reflects credit both on the author and on the company; he does such editing and/or rewriting as are necessary to effect this.
(7) He handles or expedites the clearance of articles through the publication committee, to obtain management approval.
(8) He prepares the editorials and nonassigned text material.
(9) He may ghost write contributions for company executives.

The editor also has a number of other duties, some at the nuts-and-bolts level, which he either performs himself or at least supervises. These include:

(1) Choosing typography, format, and method of printing.
(2) Handling relations with the printer.
(3) Drawing up and adhering to a budget.
(4) Laying out or supervising the layout of each issue, including fitting copy and illustrations to the available space.
(5) Writing introductory paragraphs for feature articles, to catch reader interest.
(6) Supervising distribution of each issue.

This chapter has made no attempt to cover all the functions of editors of company-sponsored publications, or to dwell on each in exhaustive detail. However, one fact does seem clear—that the editor has to make many decisions, the more important of which require a thorough understanding of his publication's basic purpose and specific objectives. A publication whose purpose and objectives are only vaguely defined is like a ship without a compass—and is about as unlikely to make its point as the ship is to make its port.

REFERENCES

1. Anon., "Operation Tapemeasure," New York, International Council of Industrial Editors, 1956.
2. Breth, R. D., "Building Editorial Balance Into Company Publications," *Management Review*, **41**, 491-93 (1952).

Chapter Five

Editing from the Patent Standpoint*

When a technical document is submitted for editing from the patent standpoint, there is at once implicit recognition that some degree of commercial interest exists and that the research results upon which the document is based represent a potential property value. This editing must therefore have as an uppermost consideration the effect of the document's publication on the acquisition of a property right, usually represented by an issued patent, in addition to the conventional considerations of content, manner of presentation, and value of the information.

Accordingly, the patent editor must first consider the advisability and propriety of publication. If he determines that this should not take place, no matter how the text could be altered, his editorial work ends with the citation of his reasons. On the other hand, if he is satisfied (or establishes by consulting others) that the documentation's publication would not be expected to interfere with acquiring patent rights, the editor then usually turns to the timing of the publication and to a detailed consideration of the

* An anonymous contribution.

document's content. With respect to the latter, the patent editor is not primarily concerned with grammar, rhetoric, or phraseology, but rather with the actual meaning and implications conveyed by the text.

In considering in this chapter whether publication is permissible or desirable from the patent standpoint, it is assumed that no over-all policy against publication exists and that each publication is to be considered on its own merits. It is essential, also, that the patent editor enjoy the confidence of many departments of his organization, and that he have at least a general familiarity with the art to which the subject pertains and with the activities of the principal competitors in the field. He can then often resolve questions of propriety in disclosing research results, from the standpoint of acquiring patents, by judicious editing—by changes in context or manner of presentation.

PUBLICATION DURING PENDENCY OF PATENT APPLICATIONS

In this work the editor bears in mind that a publication which is printed during the pendency of a patent application normally can be expected to come to the attention of the appropriate examiner in the Patent Office. It is only natural for the latter to compare the context of such an article with the application upon which he is determining patentability.

As is well known, the normal prosecution of a patent application requires considerable argument on the part of the patent solicitor to convince the examiner that the claims in the application actually constitute patentable invention. Among the many considerations are the scope of the claims, the distinction between that which is claimed and the prior art, the element of unpredictability, the detail in which the generic concept is described, the amount of data apparent in the application, and the positive tone of the disclosure with respect to operability and utility of the invention, right up to the limits claimed. These, then, are also among the factors considered in editing technical publications from a patent standpoint.

If differences of presentation in the proposed publication and related pending applications cannot be satisfactorily resolved, then it is usually concluded that publication is undesirable, at least until the corresponding patent issues. Moreover, there still may be a question about publication even after the patent issues, for statements of the inventor (or others in the organization of the inventor's assignee) that place the facts unintentionally in the wrong light may well be taken as a repudiation of the arguments presented during the prosecution of the application and accordingly may cast a cloud on the validity of the patent.

PUBLICATION BEFORE FILING PATENT APPLICATIONS

One of the essential parts of a patent application is the oath of the inventor with respect to his belief that he is entitled to a patent and has conformed to certain basic requirements. Thus, the law states in part, "A person shall be entitled to a patent . . . unless the invention was . . . described in a printed publication in this or a foreign country . . . , more than one year prior to the data of the application for patent in the United States. . . ." (U.S. Code, Title 35, Sec. 102). By publication, of course, is meant his own as well as others.

It is not unusual for circumstances to dictate the publication of research results prior to filing patent applications covering them. In general this practice contains inherent dangers that the patent attorney prefers to avoid. As mentioned above, if such publication occurs more than one year before filing an application (that is, beyond the grace period allowed by statute), there is absolutely nothing that can then be done to obtain a valid patent. From this it can be seen that cooperation is essential between those responsible for developing both the patent and the research program. In particular, the editor must satisfy himself that appropriate patent applications have been filed or that, if not, they can be filed within one year from the proposed effective date of publication.

In most organizations where acquiring patents is a significant operation in the corporate structure, the patent department carries a veto authority on publications that is at least equal to that of other managerial divisions, as a practical matter. This responsibility can quite obviously cause friction and hard feeling between the publishing group and the patent group. Resolution of differences is best considered a long-range problem, and not limited to any particular publication. This goal can be accomplished most satisfactorily by a program of education, aimed at informing each group of the facts behind the problem.

Furthermore, the realization that patent acquisition is a means of publication that can enhance a scientist's professional stature often reduces the sting that such veto power can otherwise carry. It follows, of course, that the patent attorney is charged with the responsibility of achieving high professional stature when called upon to act as the scientist's ghost writer if the right of other publication is denied. The great majority of writers of patent specifications are aware of and accept this responsibility.

EDITING FOR CONTENT

Having established that publication is desirable—or at least not undesirable—and having approved its timing so as not to impose a statutory bar against acquiring a patent, the patent-minded editor then looks to the actual content of the publication to determine what effect its statements could have upon corresponding or related patent applications. He gives particular attention to inadvertent inclusion of vague or misleading statements.

Incidentally, this discussion also applies to the content of technical publications on which there is no patent application, for such publications could conceivably have a deleterious effect on other pending applications or even on the validity of issued patents. In the ultimate case, the patent department routinely reviews and edits all technical material for outside distribution—literature articles, sales brochures, fact sheets, and advertising copy.

Method of Presentation

In considering the actual content of a technical release of any kind, editors are often faced with an apparent conflict in viewpoint on the method of presentation among research, sales, manufacturing, and patent groups. It is probably axiomatic that the patent attorney always seeks to disclose and claim an invention in the broadest terms that are possible within the restrictions imposed by the prior art. It is also frequently true that the professional scientist is reluctant to disclose in a publication any more than the specific results obtained from his laboratory work.

While theorizing or hypothesizing often form the very heart of a technical publication, speculation is infrequent therein on the implications of specific research results. The patent attorney, on the other hand, operating within the restrictions of the statutes and recognizing the competitive nature of most research programs, must file patent applications of sufficient breadth to cover the invention clearly as a generic concept within a relatively brief interval from the initial, sometimes-specific discovery made in the laboratory. In other words, the scope of an invention in the patent sense is usually broader (and quite properly so) than the scientific basis or laboratory data as of the time the application is filed. Inevitably there is an area where the attorney will rely on predictions of scope and operability resulting from the scientist's knowledge and the limitations of the prior art. Such aspects may comprise reasonable technical prediction that is evident to the expert as soon as the core of the invention is delineated.

The experimental details usually found in a technical publication may also be found in the pending applications of the same author. The patent editor must therefore compare these data in each instance. A typical conflict arises when a result is reported in the publication as being, from the scientist's viewpoint, somewhat less than desirable. Yet the patent laws recognize that such results may be a sufficient advance in the art to merit a patent. In editing the language of the publication, therefore, it is then necessary to convince the author that although in his scientific opinion his results did not reach the epitome of perfection, the very fact that he considers them sufficiently noteworthy to publish them under his name makes it obvious that it should not be necessary or appropriate to include derogatory, apologetic, or perhaps regretful remarks in discussing them.

The patent application corresponding to the publication may contain patentable claims that are considerably broader than the specific experimental data reported in the publication, an increased breadth that may be desirable from the patent standpoint of the author's employer. These claims may well encompass the ultimate commercial species while the initial laboratory proof, quite understandably, was obtained by using expensive raw materials, because of their traditional adaptability to laboratory investigations. Indeed, had the laboratory work extended out to the limits of the scope of the claims, it may well be surmised that somewhere within this framework results that are scientifically outstanding would have been obtained. In such instances the patent editor must be on the alert for remarks in the publication that would positively establish that the research is limited to the aspects being published or that no conclusions can be drawn with respect to what might be obtained if the results were extended.

Utility

In order for an invention to be patentable it must, among other things, have utility. Furthermore, if the invention is being made in a crowded art, its usefulness must be unexpected or at least a distinct advance over the same utility shown in the prior art. Research publications are frequently not concerned with utility, often being merely reports of laboratory results or of the preparation of series of chemical compounds. Such publications usually do not involve great problems in editing from the patent standpoint.

Where utility is disclosed, however, care must be taken in considering the effect of such disclosure on corresponding patent applications. A new chemical or class of chemicals may be patentable if it unpredictably is useful in an art to which analogous compounds are inapplicable. However, if the only known utility is the same as that of prior-art cousins, to which it shows up second best, it is probably not patentable on this basis alone. Nevertheless, the new compounds may be valuable in the future because of secondary considerations that are visible to the patent solicitor in establishing patentability but which perhaps do not constitute acceptable subject matter for a technical publication.

Such a situation requires close cooperation and understanding between the author and the patent editor. On the other hand, if it is properly presented, the published description of a utility which enhances patentability will have considerable weight in establishing to the examiner's satisfaction that this utility is indeed an advance in the art, and can frequently eliminate the necessity of presenting such information by affidavit.

Novelty

Technical publications frequently describe results obtained by others as a necessary background for understanding the results being published. In the United States, patents are granted for inventions, not for research results. While no one has successfully defined the term "invention" in positive terms, case law is replete with descriptions of things that an invention is not. One thing that is not invention is to take something that is known before and conduct research on a logical extension.

This fact may not appeal to the scientist, but it has much to recommend it. As one result, U. S. patents stand as worthwhile technical advances in the great majority of cases. This test of patentability recognizes that patents are not granted for mere routineering. Accordingly, the scientist can take considerable pride in the publication of his research in this form.

It is important, however, that the scientist recognize the problem as such, so that a satisfactory meeting of the minds with the patent editor can be established. For example, the author, if he is a chemist, may find nothing unusual in experimenting with a process involving a *reactive* halogen by substituting for it an *unreactive* halogen, say, chlorine. In his mind he may have theoretical or practical grounds for predicting that both would be operable. However, there is still room for disagreement in advance that the desired result will be achieved, or it may be that only the scientist involved has the knowledge which permits his positive speculation. Ordinarily, to an impartial observer this predictability has only a concrete basis in hindsight—that is, after the novel disclosure is made and compared with what has gone before.

Regardless of how the obviousness is measured, such a so-called obvious extension of the art could easily be presented in such a light that a valuable patent could be lost. While the patent editor would not insist that the technical author describe an advance of this type in such glowing terms as "surprising" or "unpredictable," he may well request exclusion of remarks that directly compare the results with what had gone before. A mere inclusion of the prior art in the usual form of a footnote will ordinarily do no harm, however, for the patent examiner in his search will ordinarily be aware of the same reference.

Valid patents under the existing patent laws are frequently obtained for minor improvements—minor, that is, in the scientific sense. Frequently, however, such patents are of extreme commercial significance. It might be fatal to patent protection if the inventor in his capacity as an author disclosed the prior processes in a publication and then described his patented invention as an obvious and logical extension, further stating that in his opinion the two processes were equivalent. The problem and conflict here undoubtedly arise from the inherent and overly modest attitude with which a scientist is prone to view his own accomplishments. It should always be recognized that the well-informed scientist is comparing his initial efforts and prior knowledge. When the discovery is finally made, it may be almost anticlimactic to the inventor. This is not so to the patent examiner, however, if his first contact with the discovery is a full-blown disclosure of a new and distinct result.

Operability

The question of operability is one in which many misconceptions occur. It is perfectly proper under existing patent law to claim material beyond the scopy of the specific laboratory data. In the reverse situation where the patent attorney is submitting his application to technical editing by the inventor, such extension is frequently questioned; the inventor, as a professional man, may even be reluctant to sign the patent application. This conflict of opinion can usually be resolved by education, and by impressing upon the inventor that he is entitled as an expert in the field to predict the logical extensions of his basic invention. It should also be borne in mind that the oath which the applicant makes in filing a patent application refers only to his belief that he is the first inventor and to his belief that he is entitled to a patent. The oath does not allege that the inventor necessarily believes that every statement in the application, were it checked out scientifically, would be completely accurate.

This is most important, for the species first tested in the laboratory, for quite logical reasons, is frequently not the species which would result in the commercial process. However, circumstances may make it necessary to file the patent application before the ultimate commercial species is actually reduced to practice. Indeed, at the time of filing it may be impossible to predict which of several disclosed species would be commercial.

In the reverse situation where the technical author is publishing a description of the first discovered species, and is limiting his publication to it, the editor must assure himself that there is nothing in the publication which can be construed as an opinion that these results could not be extended. It has been found in the majority of instances that such statements are largely opinion and that experts can usually be found who would quite honestly predict the contrary. In the absence of positive experimental evidence, therefore, the editor is on firm ground in requiring the exclusion of negative statements that bear on the operability of a process or product inside the limits of what is desired to be claimed in a pending application. It is worth repeating here again that the author or applicant should consider himself an expert in the field and should realize that he is entitled in this capacity to predict extensions that are beyond his experimental results.

* * *

The most desirable situation to achieve in editing technical publications from a patent standpoint is one of cooperation based on the recognition that the over-all goal is the same, regardless of the viewpoint. By establishing such a relationship and an appreciation of the problems of each group, watchdogging of publications can largely be eliminated. This results in fewer revisions of drafted publications after review by the patent department, and of course lessens the work of the editor. Moreover, if authors are familiar with the patent applications on file or in preparation at the time that they prepare specific publications, problems are considerably minimized. Indeed, a well-written patent application may well provide useful material in the preparation of publications on the same subject.

Facts are facts, but the way in which they are presented has real meaning in editing from the patent standpoint. Such editing is no game, played between inventor and patent examiner, but is instead a serious effort aimed at approving the release of publications that protect the property rights involved.

Chapter Six

Editing and Publishing Technical Abstracts for Internal Use*

MORRIS D. SCHOENGOLD, B. H. WEIL,
and MARY ANN MENTO
Technical Information Division
Esso Research and Engineering Company
Linden, New Jersey

The type of technical abstracts discussed in this chapter is that used in the internal bulletins of abstracts prepared for the technical personnel of a research or producing organization. Thus we are automatically excluding the editing of abstracts prepared for a full-fledged secondary journal such as *Chemical Abstracts,* about which much has already been written.

Organizations publish their own bulletins of abstracts to

* This chapter is based on two ACS Division of Chemical Literature papers, "Editing Technical Abstracts," by Morris D. Schoengold, presented in New York City in September, 1957, and "Publishing Modern Abstracts Bulletins," by B. H. Weil, Morris D. Schoengold, and Mary Ann Mento, presented in Pittsburgh, Pennsylvania, in January, 1958.

serve the specific needs of their technical staffs.[2, 3, 4, 5] These internal bulletins are usually slanted toward more specific audiences than are the major secondary journals published by scientific and technical societies. They are used instead of these journals for one or more of several reasons:

(1) The need for abstracts on a combination of subjects and fields not covered by one or two secondary journals.

(2) The need for complete coverage of applied journals as well as scientific ones.

(3) The need for more rapid publication.

(4) The need for including abstracts of internal reports.

(5) The relative economy when large numbers of bulletins are needed, as against buying multiple copies of a number of secondary journals.

(6) The absence of secondary journals which adequately cover the fields involved.

The abstracts prepared for an internal audience are preferably written by a specially trained abstracting staff, or they may be contributed by laboratory personnel on a part-time basis. The main function of these abstracts is to relate what is new in the literature to the technical program of the organization. As an auxiliary function in some organizations, they also serve later as a searching tool. This latter advantage is obtained if the abstracts are filed in such a manner as to make them retrievable.

Obviously there is no point in writing abstracts which cannot be used to maximum advantage. In order to achieve this, the abstracts must be properly edited and published.

BASIC EDITORIAL FUNCTIONS

One of the main functions of an editor of abstracts is to keep the abstracting staff well informed of company interests, toward which abstracts written for internal consumption should be slanted. This awareness may be effected in several ways. The editor must himself be thoroughly informed on the company's research program. This may be achieved by attending meetings, by reading company reports, by personal contact with the research or operating personnel, and sometimes even by judicious eavesdropping. Another valuable guide to the editor is the type of requests for information that are made of the library. In the ideal situation, the editor always attends planning meetings at which programs are mapped out. The information thus obtained must be passed along to the abstracting staff.

Another function of an editor is to train new abstracters. This subject has been covered in a recent booklet,[1] so it is not necessary to belabor the point except to emphasize its importance.

The abstracting staff should be well trained in the primary field of the company's interest, without being top-heavy. For example, a chemical company naturally requires chemists on its abstracting staff, but in most cases aspects of engineering and electronics are also important to it, so engineers and physicists may be employed. It is often the responsibility of an editor of abstracts to see that his staff of abstracters is well-balanced from the aspects of both subjects and the languages they can read.

Since one of the functions of an abstracting staff in a technical organization is to select documents that should be of interest to the research personnel and to boil down the pertinent information contained therein, it is usually up to the editor to make certain that his abstracters do not spend time on unnecessary material—and by the same token do not pass up anything of potential interest.

It is common knowledge that abstracts are not easy to read. Thus it is up to the editor to see that they are made as palatable as possible. Methods of doing this will be discussed later in this chapter. However, it should be borne in mind that short, complete sentences are important, as well as good English. The editor should read all the abstracts written by his staff to see that they conform to acceptable standards.

Actually, there are no tricks or gimmicks in editing technical abstracts. It is a matter of hard work and common sense. The editor must know what information the research or operating personnel needs to perform its functions. He must be alert at all times to the significance of the information that comes to his attention, and be able to recognize the relationship of what he reads to what he knows of his company's technical programs. And he must be capable of making quick decisions because, with the vast amount of literature being published today, he does not have the time to mull over every article that crosses his desk.

As a matter of fact, an editor need only do a minimum of actual editing if his staff is well-trained, well-informed on company interests, and well-balanced. In cases where this is not so, however, it is up to the editor to check the abstracts written by his staff for accuracy and completeness. It is important that the abstract be accurate, that it does not misinterpret the original document, and that it says no more than the original. Interpretations by the abstracting staff, if any, should be clearly delineated; actually, only an expert in the field and one who is very well informed concerning the company's interests should interpret any document.

If the editor can determine how the research person reads, what he expects from an abstract, how he uses it, and the best method of presenting the information, he is well on his way toward being a good editor. Answers to these questions may be obtained by personal contacts, surveys, and follow-ups.

STYLE OF ABSTRACTS

At Esso Research we have made several studies in order to improve the readability of the abstracts bulletins we publish. Our basic premise has been that these abstracts bulletins are prepared for reading, prior to any other use. Thus it appeared logical to write our abstracts in such a way as to make this reading easier.

Several changes have therefore been inaugurated. Each abstract is begun with a carefully selected topic sentence in which the salient idea is placed first; hence it is susceptible to being underlined or set in bold-face type (see Figure 6.1). Bibliographic data appear at the end of the abstract.

> The auto-oxidation of a mixture of isopropenylbenzene and an organic acid such as isobutyric, acetic, or stearic acid, is accompanied by decarboxylation of the acid. The rate of hydrocarbon oxidation is accelerated about 4 times in the presence of the added acid. In turn, this induces the decarboxylation reaction.
>
> "Conjoint Decarboxylation in the Auto-Oxidation of Isopropenylbenzene in Mixture with Fatty Acids."-N.I.Mitskevich, T.I.Soroko, and B.V. Erofeev(Doklady Akad.Nauk, v.115, No.1:103-06

Figure 6.1. A technical abstract starting with a topic sentence (IBM typed).

Placing the bibliographic data last obligates the abstracts writer to prepare his own topic sentence. This may even be prior use of the title of the document, if it states the subject appropriately, but more often the topic sentence can be phrased in words that lead smoothly into subsequent and supporting sentences. Abstracters now look not only for the subject of the document, but also for what is new about its message, combining both in a sentence that places the news first.

This practice involves little that is radical, but it does break with tradition. The text of most abstracts has usually begun with a topic sentence, but in practice these sentences have often been continuations in part of the more or less descriptive document titles although separated from them by bibliographical data. Or the abstracts writer, intimidated by a good title, has sometimes plunged directly into the supporting details, thus requiring the reader to glance back at the title before going ahead.

Actually, this "new" style of abstracts writing is intended to make the abstracts bulletins more like journals to which the technical reader has become accustomed, especially those which print pages of pithy "news-notes." In these, the topic thought appears in bold-face type at the start of each item, either to trigger his interest into reading further or to signal him to pass quickly to the next item. Thus, even the nonlibrary-oriented reader of such abstracts bulletins can glance quickly at each page, noting items of importance to himself and passing over those which are not.

Naturally, there is more to writing abstracts than the use of topic sentences. Details of this work have been discussed adequately by others, notably the editors of *Chemical Abstracts*.[1,6] Moreover, any discussion of informative vs. indicative abstracts, use of brief annotations, etc., is beyond the scope of this discussion, except to note that informative abstracts are usually the most popular and useful.

There is nothing new in writing any type of abstracts that in any way affects the need for conciseness, technical accuracy, and bibliographical completeness.

ARRANGING THE ABSTRACTS

It is common knowledge that the volume of literature being published today is increasing at an exponential rate. Most bulletins of abstracts, therefore, now contain at least several pages and may well be expected to grow in size. This fact, in turn, makes it desirable to divide them into logical sections for easy use by readers. To do this to the satisfaction of a wide technical audience, containing a number of specialists, is not an easy task.

To begin with, many readers seem instinctively to resent receiving a bulletin containing a large number of abstracts. The average reader knows full well that many abstracts are not pertinent to his work, and that he is not expected to read them. He knows that section headings have been chosen to guide him to the abstracts that are pertinent, and that the other headings and abstracts are intended for his colleagues. He realizes that there are economies in grouping subjects pertinent to his organization's work in a single bulletin, or a limited number of bulletins, and that each interested reader is expected to exercise discretion in how he uses his copies.

Unfortunately, many readers still continue to react adversely to multiple-paged bulletins, perhaps from the frustrated point of view that they would really like to keep up with everything pertinent to their field or organization, but know that they do not have the time. Going on from this reaction, they may feel insecure in their ability to locate the abstracts intended for them and then, obliquely, fail to place confidence in the completeness of coverage of the bulletins. Also, they sometimes place the multipaged bulletin of abstracts in the same category with other multipaged documents—to be read if and when time permits. Since many organizations publish their bulletins weekly, this attitude can and does get the reader into trouble. Bulletins stack up on his desk, and he either reads several issues at a time or discards batches of them. Less frequent publication would not be the answer, however, since that would compound the problem as regards size of the bulletins.

Some organizations, confronted with the fact that their bulletins of abstracts were not being read, have cut their size drastically by selecting only those abstracts that seem of outstanding importance, and by limiting their bulletins to a specific maximum size. These moves often fail to give the technical reader the complete coverage that he desires, and hence do not aid him materially in cutting down his personal reading.

Esso Research has selected section headings that are in the language of the readers, particularly in the terminology employed in a specific division, and has even personalized such headings to the extent of titling a section "Butyl and Other Elastomers" instead of, simply, "Elastomers." Of more importance, probably, is our use of amplified subject headings, carefully phrased to give the reader more confidence in the subjects included under a given heading. One such heading is shown in Figure 6.2.

THERMAL CONVERSIONS

Theory, methods, problems, and equipment. Includes visbreaking, thermal and steam cracking, and thermal reforming (see also COKING).

Figure 6.2. A descriptive heading for an abstracts-bulletin section.

Subject headings in bulletins, incidentally, cannot and need not be as specific as those in indexing or classification systems, and it is often possible to group related material advantageously. However, this grouping may work something of a hardship on certain readers, especially those who are interested in very specific subjects. That is why there has been such a clamor for subject indexes to issues of *Chemical Abstracts*. Subject indexes have also been considered for internal bulletins of abstracts.

The best possible section headings, therefore, are those which are broad enough to encompass related material, without grouping too many abstracts under a given heading, yet specific enough to enable the bulletin reader to identify his specialties, preferably in language which he himself would have used.

Naturally, while headings may be mutually exclusive, documents often cover several subjects. Many bulletins ignore this, leaving it to the reader to scan all sections which might contain pertinent abstracts. Others place appropriate cross references at the ends of the various sections.

As discussed later, some organizations are avoiding excessive size in their abstract bulletins by publishing several separate bulletins, divided according to type of document (literature, patents, etc.) or subject. The latter method presents several problems unless the organization's work is in several fields which can be sharply differentiated. A related approach has been to select peripheral areas, such as applied mathematics, corrosion, safety, etc., and to publish separate bulletins on these. Such bulletins are then made available to any reader of the former bulletins who desires them, and gain new readers who previously were unwilling to look for these subjects in broader-scope bulletins.

"HIGHLIGHTS" SECTIONS

A "highlights page" or a "highlights section" at the beginning of each bulletin is another device now used to interest and serve the readers of abstracts bulletins. These either present the salient information about a few items of unusual importance to the company's work, augmented with brief comments, or they are paragraphs which briefly review all the abstracts contained in the bulletin, grouped according to subject.

Manufacturing

LITERATURE

TECHNICAL INFORMATION DIVISION BULLETIN

ESSO RESEARCH AND ENGINEERING COMPANY

Volume 38 December 11, 1957 Bulletin I, No. 50

Highlights . . .

Fatty acids accelerate the oxidation rate for isopropenylbenzene by a factor of four. The fatty acid, such as acetic or stearic acid, decarboxylates during the process, according to Russian workers (1005b).

Figure 6.3. A "highlights" item and masthead for an abstracts bulletin

Esso Research employs a highlights page (see Figure 6.3) which is a single sheet that contains basic information from and about several of the most important references, and which is also the cover of the issue. This page is intended to serve at least two purposes. One is to give at least some important information to the reader (such as the executive) who will not devote time to reading more. The other is to persuade previously reluctant readers to use the bulletins properly—at least to open them for further information on items of real interest. Such pages also stimulate the regular readers of the bulletin by guiding them quickly to items of outstanding interest.

Highlights pages, of course, are first cousins to the sections in modern technical journals which present the salient information in the more important articles in terms of reader interest, in order to impart at least some facts to all readers or, more important, to invite the interested reader to turn to the full account.

Items for the highlights page of an abstract bulletin must be selected by people who are expert both in the fields involved and in the company's specific stake therein. They must be brief, but must contain enough details to verify their importance. For quick reading they should start with topic sentences, with the most salient news first. Each item must be keyed to the complete abstract, usually by a number in parentheses.

Editorial comments relating the items to the company's research program are highly desirable; however, they should be clearly distinguished from the facts presented in the reference. Incidentally, there has been increased pressure in some companies to have qualified abstracters make interpretive additions or substitutions in the abstracts themselves, in order to pinpoint matters of interest to company work.

If the highlights page is properly written, it may also be used as a separate one-page bulletin for some readers (chiefly executives) who were previously unwilling to look at the larger, more complete bulletins.

Some organizations begin each issue of a bulletin of abstracts with a highlights *section* that contains several paragraphs or pages of running summations of all or nearly all of the abstracts in the bulletin. The topic of each abstract is thus presented, so that the interested reader can refer to the full abstract for more details. Abstracts that are not of interest, even though noted in the same paragraph of the highlights section, can be disregarded without further ado.

These highlights sections are usually written in a concise yet flowing style, much like such industrial-information bulletins as Gaylor's *Technical Survey*. They must be prepared by writers who understand the importance of each abstract to the company's program and/or individual interests, and who can phrase the abstract's subject in the appropriate style. In essence, these highlights sections are running subject indexes, presented in a highly palatable fashion.

PUBLISHING

Publishing the modern bulletin of abstracts, once the style of abstracts and certain other features of the bulletin have been set, involves decisions concerning staff, editing, coverage, design, duplication (typing and printing), and distribution. Staff requirements and editing have been discussed previously.

Coverage

Internal bulletins vary widely in their coverage of types of documents. Some bulletins cover only the periodical literature, or patents, or reports and other internal publications. Some include several of these categories.

Some organizations, as mentioned, publish more than one bulletin, separating abstracts of the literature, patents, and reports, or separating the abstracts according to subphases of the organization's work or interests.

The abstracts are usually prepared by scanning the primary literature or patents, selecting pertinent items, and writing the type of abstracts decided upon. Some organizations combine this direct coverage with inclusion of selected abstracts from secondary journals. Economics—the cost of subscriptions and abstracting—is usually the deciding factor. However, timeliness—one of the chief attributes of the usual company abstracts bulletins, over and above increased selectivity—is not possible if secondary sources are used to a large extent.

It should be mentioned, incidentally, that some organizations obtain their abstracts from central abstracting services in their fields, such as the API Technical Abstracts Service in the petroleum field and the Engineering Index Service in the engineering field. These abstracts are prepared quite currently, and quite well, and subscribers to these services may use the abstracts intact, edit them if they desire, and/or intersperse them with abstracts which they prepare themselves. The API Technical Abstracts Service, moreover, publishes its abstracts in bulletin form, and some companies in the petroleum field distribute these in place of their own bulletins.

Many companies, in selecting material to be abstracted, differentiate between abstracting formal documents—technical papers, books, reports, patents, and preprints—and more transitory material such as news-notes and advertisements. Thus, the typical bulletin of abstracts provides the technical man with a survey of the more formal parts of many journals which he might otherwise feel necessary to cover, but often it does not "keep him up with the news."

In one sense, this is not particularly surprising. Most companies do not intend that their bulletins of abstracts should stop their technical people from reading journals to keep abreast of the latest "news" in the fields of their interests. The chief purpose of such bulletins is to relieve the technical man from having to cover many journals for fear that he may miss something of basic importance.

A few companies publish summaries of the news as parts of their bulletins of more formal abstracts, or as supplementary bulletins.[7] Others distribute publications which accomplish the same result, such as Ethyl Corporation's *Petroleum Refining Developments.*

Each organization must decide for itself how far it can go toward the desired completeness of coverage. A balance must be struck somewhere among expense, time, excessive length, and the confidence of readers in the completeness of bulletin coverage.

Design

If an organization publishes several different bulletins of abstracts, it must take care to differentiate visually among them. Even if the bulletins are published as parts of a series, under a series title, the specific title of each bulletin should stand out clearly. Other devices used to differentiate among bulletins include first pages of distinctive colors and emblematic drawings or pictures.

Most bulletins are printed on white paper. Some organizations, however, have found that the use of ivory paper reduces glare with certain typewriter styles, especially if the duplication process occasionally yields pages that are darker than desired. Use of ivory paper may also add to the attractiveness of the bulletins.

Duplicating

Most of the secondary journals published by societies employ letterpress printing, or offset printing from typeset material. Few internal bulletins of abstracts have been able to afford the cost of such publication, which represents a disproportionate expense when the bulletins go to a limited number of readers. Accordingly, the hallmark of most internal bulletins of abstracts is the use of typed material. In recent years, however, quality of appearance has been improved by the use of typewriters with proportional type faces, and even occasionally by the use of special typewriters ("Vari-Typers") which simulate typeset composition more closely through the availability on the same machine of types of different sizes in light, bold, and italic styles.

The average technical reader, of course, equates quality of appearance with typeset columns that have straight right edges, and unconsciously gives a poorer rating to typed material, although he is accustomed to reports prepared in this manner. Accordingly, he sometimes unconsciously low-rates a typed bulletin of abstracts, a factor which must be taken into account when, even with restricted budgets, typewriters are selected and methods of duplication are chosen.

When "regular" typewriters are all that are available, care must be taken to achieve evenness of appearance, fairly regular right edges, and full-sized reproduction. If any factor of "prestige" is to be achieved, offset printing is almost a must. High-quality offset duplication from "paper" plates or mimeographing is sometimes acceptable, but hectographing yields a bulletin which is more difficult to read and has the unintentional appearance of haste.

Quite acceptable results (see Figure 6.1) can be obtained by the use of typewriters with a proportional type face, such as the IBM "Executive." Small photo-reductions are permissible if line lengths are kept short (as in bulletins which employ two columns per page). Again, photo-offset printing is the preferred method of reproduction; offset printing from "paper" plates should not be ruled out, however, when limited editions are required. Mimeographing is possible, but, again, hectographing yields undesirable results.

When Vari-Typing is employed, the most satisfactory method for taking advantage of the improved appearance is photo-offset printing or one of its variants, such as XeroX-offset duplication. Neither mimeographing nor hectographing yields desirable results.

Distributing

Since bulletins of technical abstracts are published in order to place current information in the hands of those who can use it, enlightened organizations see to it that these bulletins reach all who desire copies. Moreover, since publication of these bulletins usually places emphasis on speedy reporting of technical developments, individual copies are normally provided.

Physical distribution involves little that is unique. Each issue of the bulletins must be sent out promptly. New "subscribers" must be added quickly, and other changes must be made without delay. Bulletin receivers, like all subscribers, can be unduly upset by apparent trifles.

SIMULTANEOUS PREPARATION OF FILE CARDS

A problem that confronts many technical-information groups is the economic need to produce both abstracts bulletins and file cards from a single typing. In some cases the solution chosen has been detrimental to the bulletin reader; i.e., has required the preparation of conventional, bibliographically-oriented abstracts.

This might well be taken to mean that these two uses of abstracts—bulletins and cards—ought not to be combined. If economics demand this combination, however, there are ways in which the disadvantages can be minimized.

One solution that has been advanced involves spacing the abstracts appropriately for cards; running sheets or card stock on one side only, along with the bulletins; and then cutting these sheets apart or separating them along preperforated lines. This technique has the disadvantage of printing only a few abstracts per bulletin page, and some bulletins printed in this manner—although they permit readers to prepare their own card files—are so bulky that few readers care to handle them.

Another method, surprisingly inexpensive when only a few copies of each abstract are needed, involves printing extra copies of the bulletins (preferably on one side of the sheets), cutting them apart, and pasting each abstract on a card. Additional bibliographic indicia can be typed on each card if desired.

A variant of this method can be employed if more than

a few copies are needed of each abstract. Copies of the abstracts are pasted on sheets (several to the sheet), additional indicia are added if desired, and duplicating is done by XeroX-offset printing.

If offset printing of the bulletins is done from direct-typed "sticky-back" masters, abstracts on these can be cut apart after the bulletins have been run and then respaced for the running of the cards by tacking them to the surface of regular "paper" plates. If additional indicia are desired for the card files, they can be typed on other sticky-back masters, cut from them, and tacked down in the appropriate places along with the abstracts. If the bulletins are photo-offset printed initially, rather than run from "paper" plates, it is possible to print sticky-back plates from the photo-offset plates for cutting apart and later running of cards in the above manner.

These and other methods are certainly cheaper than retyping and proofreading a set of masters specifically for cards. If the bulletins contain abstracts beginning with topic sentences and with bibliographic data at the end, it may be desirable to cut the abstract apart and move the bibliographic data to the top before running the cards. This facilitates use of the cards in reference searching for authors, patentees, dates, etc. On the other hand, if whatever indexing or classification is desired is added in the margins of the card before running, then later reading of segments of the card file may be facilitated by leaving the abstracts with topic sentences first.

Incidentally, special care must be taken in writing abstracts that are used simultaneously for current-information bulletins and for card files. Indexing or classification, if it is at all deep, may yield cards that turn up later as apparent "false drops" if each abstract has not been properly written to show the pertinency of other subjects than the one shown in the topic sentence.

Internal bulletins of abstracts can be highly useful documents. Each organization, however, must decide for itself whether it really needs such bulletins, how extensive they should be, how they should be slanted, and how they should be distributed. Moreover, since reader needs and habits are not static, occasional surveys should be made to determine whether changes in the bulletins might improve their utility.

Time is now really "of the essence." Abstracts bulletins which have been properly prepared can be very valuable tools for the technical man.

REFERENCES

1. Crane, E. J., "The Training of Chemists for Abstracting and Indexing," pp. 16-21 in "Training of Literature Chemists," Washington, D. C., Am. Chem. Soc., Advances in Chemistry Series No. 17, 1956.

2. Fleischer, M., and Hooker, M., "Variation in Content of Abstracts According to Use," *J. Chem. Educ.*, **33**, 27-32 (1956).

3. Ford, Genevieve, "Use of Library Bulletins," Paper presented before the Metals Division, Special Libraries Assn., Detroit, 1955.

4. Jackson, Lucille, "Some Observations on Fifty Technical Library Bulletins," *Special Libraries*, **44**, 366-69 (1953).

5. Sewell, Winifred, "Problems of Issuing a Company Pharmaceutical Abstract Bulletin," *Special Libraries*, **45**, 63-66 (1954).

6. Staff, "Directions for Abstractors and Section Editors of Chemical Abstracts," Columbus, Ohio, Am. Chem. Soc., 1952.

7. Weil, B. H., "The Literature Summary," *J. Chem. Educ.*, **28**, 572-75 (1951).

Chapter Seven

Training the Internal Technical Editor*

LUCIE D. CONNOLLY
Chief of Technical Editions
Los Alamos Scientific Laboratory
University of California
Los Alamos, New Mexico

Unprecedented scientific advances and the resultant need for large-scale technical communication have necessitated substantial investments in technical publishing. But the success of a company's reporting program is not entirely a function of the budget. It depends in large measure on the caliber of the editorial staff employed to produce the reports and publications. This, in turn, depends not only on the ability of the editors, but also on the quality and amount of guidance provided by management.

The problem of training editors to work in today's diverse, complex fields of research and development has appeared formidable to many companies. Basically this is because technical reporting as we know it today is a relatively new

* Based on a paper presented before the Society of Technical Writers and Editors, New York, November, 1957.

field that is still in the development stage and is therefore difficult to view in its proper perspective. Lacking experience in this area, management does not always realize that, given thorough indoctrination and training under the proper conditions, editors can assume most of the burden of producing literature that is a credit to the company.

This chapter describes a system whereby management and its editorial staff can operate in a climate of mutual understanding at a high level of efficiency. Three things are considered: what management needs in its editorial staff, what the editors can expect of administration, and how the objectives of both can be achieved.

MANAGEMENT'S NEEDS

First, let us see what management wants and needs in an editorial staff. In most organizations, research supervisors are not in a position to make an exhaustive review of manuscripts submitted for publication. Although they must assure that the work reported is significant and technically sound, it is poor economy to have executives spend time on editorial and production problems such as errors in accuracy, flaws in logic and presentation, and appropriate methods of document duplication. These are the functions of an editorial group. Therefore, a company should rely on its editors to work with scientists and engineers to convert rough copy into a manual or report that reflects favorably on its standards and that meets the requirements of the audience for which it is intended, whether it be lay or technical. It should expect these editors to produce reports at varying levels of quality commensurate with their value, expected distribution, and anticipated lifetime.

MANAGEMENT'S OBLIGATIONS

Next we should consider management's obligations to its technical editors. It is of prime importance that those who have earned senior status be accepted as professionals in their field, ranking with other research and production personnel and accorded the same considerations and prerogatives. Also, regardless of their position on an organization chart, technical editors should be recognized as staff, or administrative, rather than research and production personnel. This is too seldom done but is very important. Editors are agents of management. Their job is to produce literature that represents the thinking and standards of the company, not just a particular department. They must fit departmental work into a company-wide production system.

The channels to management must be kept open, and editors must be apprised of scientific developments and over-all policy and programming. Editors can do an adequate job of producing reports and other publications that are in accordance with the viewpoints and aims of the organization only if they are kept fully informed on its operations.

From the foregoing we can see that considerable reciprocity is needed if an editorial group is to fulfill its obligations and to operate efficiently. But it is also clear that cooperation alone is not enough. Moreover, prior education and experience in themselves cannot equip a man to handle publication of today's research with its highly advanced and complex level of knowledge and expanding frontiers.

EDITORIAL TRAINING

This brings us to the third point, the essence of which is how management can train an editorial staff so that, following established policy, it can produce a result that meets the needs of the intended audience with the quality of presentation and appearance that is warranted.

Let us study the case history of one company which has adopted a program of in-job study and training for its editors. This research organization represents a type that is growing and expanding in this country—one in which tremendous diversification demands that each professional man be a specialist and that each carry his unique load. Among these specialists is the technical editor. His educational background may include a liberal-arts major with basic courses in science, or he may be a science major with sufficient courses in the use of the English language to provide a thorough background for technical editing. As a trainee in the organization he is viewed as having talent and ability but as being far less than expert in the technical aspects of the research work.

Indoctrination

The editor's practical training starts when, as a potential senior member of the staff, he participates in an indoctrination course given to newly hired scientists and engineers. This course begins with a two-week series of orientation lectures and tours designed to acquaint new technical employees with the over-all picture of the company's work. Illustrated talks and motion pictures are presented to describe the projects of each research group, their application to a broader program, and the place of that program in the over-all operation. He participates in conducted tours of the laboratories and observes work in progress. Here the new editor learns what the reports signify, what purpose they serve, and what their potential readership is. He sees the operating scientist in his environment and becomes acquainted with his problems. This knowledge is invaluable later, when he must determine the psychological approach to the problems of individual authors.

After this preliminary indoctrination the editor attends an accelerated course in the fundamentals of physics, chemistry, and engineering as they apply to the work of the organization. This series of lectures, also a company-wide program, is calculated to acquaint the technical staff with the purposes, principles, methods, and techniques of the various fields of research. The course is given by a qualified teacher who is expert in the pertinent subjects. The editor is encouraged to investigate problems that he does not understand and to ask for explanations. Textual materials are provided for immediate use and later reference. Other literature is suggested for further study.*

The final phase of the new editor's education in the technical aspects of the company's work is a continuing one. Participation is optional and is governed by the amount of time that can be devoted to it, but the response is always enthusiastic. It involves attendance at and participation in meetings held two or three times a week. One is conducted under the auspices of management to inform the staff on major projects, proposals, and accomplishments. At times, key people from other organizations give talks on related work. The other meetings, frequently seminars, are held by groups within the company to exchange information and ideas on work in progress.

Editorial Supervision

More specific training is conducted within the editorial group under the careful guidance of an editorial supervisor. Here the new editor is taught that he cannot confine his efforts to improving grammar, punctuation, and syntax. He is expected to know the significance, purpose, and intended

* Although not directly related to its indoctrination program, the organization's contribution to further academic training is worthy of mention. Along with the rest of the professional staff, the editor is given an opportunity to take college-credit courses leading to an advanced degree in science with financial assistance from the company.

audience for each document that he handles, and to tailor it to meet a specific need. He will be editing reports and papers designed for readership ranging from the informed layman to the top-ranking scientist and engineer. Some documents will reach a limited audience; others are intended for public use. These factors determine the editorial approach that will lead to the desired end-product.

Although a measure of standardization of terminology and notation is possible, the editor soon learns that he must accommodate the editorial style to the readership. If a report concerns an abstract phase of higher mathematics, the style and vocabulary must be that of the mathematician. When he handles a training manual, the editor must work from the viewpoint of elementary, definitive presentation. Thus he must study texts and journals in a number of fields and familiarize himself with the language and style of each.

Since, in addition to the general considerations of audience and style, the editor must have some understanding of the subject to approach a report intelligently, an integral part of his training comprises review of relevant existing literature. Therefore, when a draft is assigned to an editor in training, he is given a list of authoritative books and reports to study for organization, style, and level of presentation, by means of which he can familiarize himself with the subject. This material is the basis from which he makes a critical evaluation of the draft that he is handling. Experience has shown, too, that it enables him to discuss the report with the author with some comprehension of the subject and therefore with some understanding of individual problems.

The editor is encouraged to use a personal approach in dealing with the author. Although the conventions of a particular field must be recognized and followed, he must consider the preferences of the writer as well. In other words, although his efforts may improve and correct a report or paper, he may make no arbitrary changes. He should also explain the reasons for suggestions and modifications so that the author can benefit from constructive comments when he prepares future documents.

When the technical editor has reached the point where he can speak with some authority on the editorial practices of any given field of science and technology, he finds he has acquired a following among research and production scientists and engineers. They are eager for critical assistance. He is then capable of independent performance, with supervision confined largely to matters of policy and over-all programming.

Graphic-Arts Training

The third phase of the technical editor's training is in the graphic arts. In this area he not only acts on behalf of management, but is also the author's agent. It is his job to see that the author's needs for drawings and photographs are filled. He is expected to know and choose the type and style of illustration that will convey the message adequately, and to select the method of reproduction that is suitable and economical for the quality desired and the number of copies produced. He must see that the components of the report are prepared on schedule, that the copy is suitable for reproduction, and that the material is submitted to the printer with adequate instructions. These duties are discussed in some detail in Chapter Fifteen.

To meet this responsibility to authors, and therefore to management and the reader, the editor in training is introduced to the graphic-arts staff concerned and is given sufficient training in illustration and reproduction methods to know how this work should be done to suit company's needs and those of recipients. He learns the reproduction intricacies of a 20-copy progress report with rough sketches and reproduced by Ozalid, a 500-copy formal report with finished artwork and printed by offset lithography, and a 5,000-copy handbook printed by letterpress.

He learns through instruction and observation that reports can be produced inexpensively and quickly. He studies photographic reproduction to learn what it can produce in terms of resolution of line copy, faint oscilloscope traces, and fine-grid graphs. He learns enough about letterpress printing to be able to deal with a commercial printer in specifying the company's requirements and fulfilling those of the printer.

Cost of Program

The foregoing is the essence of one organization's editorial training program. Let us now consider the time and expense required to conduct it. Inasmuch as the lectures and seminars are offered to the entire scientific staff, there is no additional cost for the attendance of the technical editors. Although the initial-productive-time loss may be three weeks, the knowledge acquired makes this insignificant in the long run because it greatly reduces editorial working time. A grasp of the basic scientific concepts enables the editor to handle reports with far greater efficiency than would otherwise be possible. Furthermore, he develops the capacity to act independently and the ability to use judgment in dealing with authors and with matters of company policy.

* * *

These training techniques have been presented as a case history. Other companies employ programs that differ in degree or detail. A recent survey of the on-the-job training of literature chemists,* including editors, listed general-

* B. H. Weil and O. E. Kurt, "On-the-Job Training of Literature Chemists in Industry," in: "Training of Literature Chemists," Washington, D. C., Amer. Chem. Soc., Advances in Chemistry Series, No. 17, 1956.

indoctrination programs and such specialized training programs as instruction manuals, citation of instructional literature, opportunity to take additional course work, encouragement to participate in the activities of appropriate societies, provision of classes or seminars on related subjects, and allowance of time for general reading. Some companies employ formal training assignments that rotate the trainee through associated sections.

Returning to the training program detailed here as a case history, one cannot advocate such a system in every organization because companies have different administrative policies and different reasons for existence. However, at least one of the three phases of training will certainly find application in a large number of organizations. The first of these—education in the aims and program of the organization—cannot be overemphasized. It is invaluable in providing the opportunity for an editor to develop to the point where his contributions result in competent, efficient, and economical production of reports and other publications.

Part II

EDITING JOURNALS

Chapter Eight

Editing of Technical-Society Journals: The Editor's Viewpoint

ALLEN D. BLISS

Assistant Editor, Journal of the American Chemical Society
Simmons College
Boston, Massachusetts

The term "technical-society journal" covers a wide variety of scientific fields, types of material, and production volume. It includes the semi-popular but nevertheless research journal, the large pure-research periodical, and the small circulation magazine in a narrow scientific field with only a dozen or more papers per issue. Regardless of field, circulation, or size, if good standards of science and literary quality are to be maintained, a journal must be guided by an interested, competent, thoughtful editor, whether he be a jack-of-all-trades near-volunteer part-timer or the chief of a well-staffed editorial office.

Technical-society-journal editors are chosen in a variety of ways and on the basis of many criteria, ranging from slight knowledge of the subject plus availability and willingness, to long connection with the field coupled with some previous editorial background. The choice of an editor for a technical-society journal usually is made by a selection committee responsible to the directors of the society, after a study and appraisal of the qualifications of eligible prospects. Particular attention is given to their skilled and thoughtful familiarity with the technical field, amplified by previous experience in dealing with people and in editorial work.

A new editor should bring with him a fair general knowledge of his society's technical area, both as to past and current research activities. He need not, in fact could not, have personal research experience in more than a very few of them, but his background of knowledge should give him an awareness of progress in all fields, and a fairly wide acquaintance with the active workers and research directors in these areas. The editor also must have (and further develop) a guiding philosophy for his job; this will be mainly a kind of subtle sixth sense which tells him when a manuscript is "right," that it lies inside his journal's field, and that it is well-planned, well-documented, well-done, well-organized, and well-written. He must have a clear mental picture of the type of science he wants to print, and the form in which it should be presented. At the same time he must have the vision to judge fairly the borderline and new-field contributions which come to him, and to expand or narrow his editorial acceptance limits in reasonable and defensible fashion.

The evolution of the technical-society editor has passed from the old-time spare-time one-man portfolio, through the "Publication Committee" stage, to the full-time editor with a large office. Modern technical editors may have assistant editors, an editorial board, secretaries, draftsmen, file clerks, library researchers, translators, copy editors, and indexers, and all the modern equipment (both typographic and photographic) that goes with them. All these seem to be necessary to record and speed the progress of hundreds of papers through the editorial process, both as to selection and final publication. Nevertheless, there are smaller technical journals of excellent quality and standards whose budgets still allow only a part-time editor, secretary, and managing or production editor.

REVIEW BY REFEREES

We need not dwell upon the virtues, etc., of the old system in which the editor himself glanced at or read each new item and accepted or rejected almost at once, on his own general knowledge. Nowadays, nearly all technical-society editors require two or three copies of each new paper and assign them to board members or other referees familiar with the special field of the study. These reviewers are expected to form a considered judgment of the paper, both as a contribution to science and as a contribution to the literature of the science (terms which are not synonymous, as every editor and reviewer knows). Having done this, the reviewer is expected to compose this judgment into a report back to the editor a report which makes clear to the editor and author the good points and faults of the manuscript, and which also offers constructive recommendations as to ways in which the paper may be improved, in case this is desirable and possible.

Reviewers for most modern technical journals do not work under any strict mandates. Their task is to judge and criticize each paper in the light of many factors, the chief of which are coverage of background literature (the prior art); the field and purpose of the research; the quality of the work and the care shown in its execution; the mode and extent of presentation of the results; the interpretation of these results; the conclusions drawn; and the organization and literary quality of the composition, plus the writer's attention to style usages, nomenclature, and symbolism. All these manifold aspects of an article are legitimate subjects for comment by a referee, and skill and thought are needed in formulating a useful, convincing, instructive report on a paper. This is especially true in case an editor uses the anonymous report system, since the referee must do his job well and not reveal his identity by word, citation, reference, or implication, or even paper watermark. Reviewers' reports usually are requested in duplicate, for record purposes and transmission, thus saving secretarial time (unless editing of the report becomes necessary).*

EDITORIAL ACTION ON REVIEWS

Fortified by one or more reports, the editor now has the task of deciding the fate of the manuscript. Favorable reports from competent referees make life easy for the editor and happy for the author, unless the editor comes upon one such as this: ". . . well-planned, well-executed, well-written, and of interest to perhaps 0.01 per cent of our readers."

Editorial gray hair and fits of depression are aggravated by those cases in which one reviewer pronounces a paper excellent and acceptable, while another reports it as verbose or poorly done, or early literature not searched (with citations to prove the point), improperly constructed apparatus or electrical hook-ups, inadequate preparation methods and imperfect purification of compounds, or erroneous calculations (again with evidence). Such problems sometimes may be resolved by additional reviewers' reports, leading to

* A more detailed account of the role of the reviewer is presented in the next chapter.

reasonable action. If the criticism is directed at the length or literary quality of the article, constructive suggestion and example may aid the writer in producing an acceptably revised paper. At times, even experimental faults may be remedied and objections met by a bit of extra work, by tests or analyses.

It is in this broad area, concerned with good papers containing minor flaws, that really constructive work by reviewers and editors is so important in the adult education of authors. Editorial reports and letters to authors help them to a better perspective on their writings, and reveal blind spots or overlooked details. It is very easy for a writer to become so familiar with and immersed in a paper that he cannot view it objectively. No one likes to be (or should be) "shown up," but every publishing American scientist will admit to having improved his published papers as a result of reviewers' careful reading and study: such details as missed references; wrong translation of idioms; too-wordy introductions; mismatched subject and predicate and tense; misspelled names (personal, corporate, trademark); over-condensed description; unnoted neat apparatus tricks; new sources of better equipment or chemicals; recently or obscurely published improvements in preparation and purification of intermediates; new analytical methods or indicators; a simpler electrical hook-up; a new regulating gadget; overlooked sources of error in observations or calculations; better methods of reporting data or of tabulation; tricks in preparing charts and graphs; the latest development in interpreting data on velocity, exchange, or rearrangement; a new application of tracer or enzyme; need we go on? The editor is judge, teacher, and research director; his editorial board a corps of lecturers and laboratory assistants. Their contribution to a paper may vary from an appreciative word of praise, a helpful hint or detail, even to a complete rough redraft of a whole paper.

DECLINING PAPERS

There are three broad causes for declining papers (rejecting is an ugly word): (1) subject matter that is definitely outside the avowed field of the journal; (2) such an extreme lack of novelty or scientific merit that the paper is unacceptable or seems incapable of satisfactory revision; and (3) the editorial conclusion, supported by adequate referee reports, that although the paper contains some good material its publication would not be in the best interest of the journal.

Few technical societies or editors subscribe to the idea that membership in a society obligates editors to accept contributions from its members. However, since rejection of a paper for one of the above reasons (often interlocked with other details) is very hard for an editor to decide justly, some further consideration of them may be worthwhile.

Good papers that fall in the first category—that are outside the journal's field—are a real editorial perplexity, and come in considerable variety. To the editor of a journal on fundamental chemistry, for instance, may come a paper which reports only a little experimental work and then presents an excellent survey of an important special phase, or a fine presentation of a teaching problem, or a complete working out of a mathematical physical problem which could be applied to chemistry but isn't. Then there come in good papers of an applied nature, usually in chemical engineering, as well as studies of closely related plant species in the light of their alkaloid or steroid content, or extensive tests of series of chemicals in growth or diseases of hamsters, or the probable relationship of certain nitrogen-sulfur derivatives in nerve or glandular response. Nearly all such contributions are carefully prepared by reputable scientists, but the editor has to ask his referees and himself, "Is this excellent manuscript the kind of chemistry we are supposed to print? Will we do this author and his colleagues a service by printing (and perhaps burying) this article now, as well as later ones on the same subjects? Should we expand into this field?" There is no set formula by which an editor can decide such problems; there are many variables concerned, and each case must be studied on the merits of the composite picture (writer, research field and its probable future, the paper itself, trends in related research, supply of papers, state of the budget, alternative outlets).

As regards the second cause for declining a paper, probably no scientific writing is totally devoid of value, other than the occasional screwball or crackpot paper which comes to every editor. In chemistry, however, there are two broad types of writing which permit quick declination: the amateurish-survey paper which may have some slight experimental tinge but does not even report good elementary chemistry, and the more-frequently-received type which is based on scanty or no coverage of previous literature, some poorly done (perhaps semi-repetitious) experimental work with crude apparatus and unproved chemicals, and which ends with sparse conclusions which may be wrong, incomplete, or valueless.

Manuscripts in the third category nearly always are special cases in which reviewers disagree among themselves and with the author. Fortunately, they are infrequent, and must be treated on an individual basis. The paper may be an extensive survey and recalculation of special field data, about which one reviewer reports "fine piece of work which will be of great value to many workers," while the next report says "not done critically, and any prospective user will wish to rely on his own recalculation." Of the next one, the first report may say ". . . excellent experimental work leading to challengingly novel conclusions," which must be paired against ". . . no fault in the experimental work, but the conclusions are not warranted by the evidence, especially in the face of other work and present theory." These papers are a real source of worry to an editor, as they seem always to involve prominent and skilled scientists. Criticism accrues to the editor whether he accepts or rejects; even in science, "chickens come home to roost."

EDITOR-REVIEWER-AUTHOR RELATIONS

The American Chemical Society's journals are outstanding examples of periodicals in pure and applied science which operate with anonymous-editorial-reviewer plans. Their present systems are the result of some 50 years of formative, trial-and-error development. There have been growing pains, and still are, but most reviewers and authors now enter whole-heartedly into the spirit of the system. However, there has been a slowly growing sentiment among referees not to remain anonymous, and among authors to suggest reviewers in their own field and others. This slowly changing atmosphere and attitude in science is jointly the result of better communication between scientists, increased development of feelings of fair play and sportsmanship, and the modern cooperative team attack on large and expensive problems.

The editorial process has its assorted by-products and side-reactions. The editor soon learns which referees are prompt or slow; which give short, snap, or cursory judgments; and which return careful, complete, constructive reports on those papers which need them. Thus he comes to know his course of action when terms expire and new board members are appointed.

A good editor comes to know his contributors and their fields, as well as his reviewers and their interests, and the skills and failings of both. Thus he is able to preserve good public relations and still achieve what is best for all. This is accomplished in many ways: blunt treatment or kid-glove handling; use of carefully edited reports to eliminate occasional picayune, annoying, or irrelevant criticisms; outlines or redrafts of unskilled, amateurish, or verbose papers; approval of more extensive presentation; tactful prevention of priority arguments or controversies; and—at all times—the preservation of the dignity of the office and the considered responsibility of its actions.

During editorial handling, a variety of minor faults in manuscripts should be found and remedied: poor or illegible typescript, lack of or substandard illustration copy, bad literary composition, unacceptable nomenclature, and deviations from good usage and office style. Much of this detail work is attended to in revisions before acceptance (reviewers are encouraged to note such things in their reports), and the rest is taken care of in the office editing of copy before it goes to the printer.

The technical-journal editor, especially if he has a managing editor located elsewhere, may not have any extensive contact with the publishing process after acceptance of papers, but a surprising amount of its detail may reappear in his office by way of questions on nomenclature; patent-date problems; matters of symbolism or spelling style; choice of method of printing tabular matter, formulas, or illustration copy; and other typographic details which affect the budget. All these questions and decisions are a part of an editor's responsibility to his directors, and of his mandate and duty to his journal's contributors and readers.

MANAGING-EDITOR ACTIVITIES

A brief survey of the activities and problems of a technical managing editor may be appropriate here, since one office may carry on all the editorial work in the case of a small journal. Every printer welcomes well-prepared and clearly edited copy, while "dirty" copy is hard and slow to set and proofread, and always causes extensive and expensive changes in proofs. Printers know this; editors find it out by experience. Technical editors should therefore approve only good, clear copy, and all material should be office-edited before it is transmitted to the printer.

This work should be done or at least supervised by someone familiar with printing-house procedures. Attention should be given to a great variety of details: uniformity of style in spelling, punctuation, abbreviations, symbolism (Greek and other characters clearly identified), bold-face and italic type, footnotes and literature citations, trademark names, centered and side heads, tabular matter and its systematic and economical arrangement, formulas and equations (their arrangement, symbolism, and column adjustment), and cut copy of all kinds (remembering that it usually is reproduced photographically) with its artistry, legibility, captions, and problems of paper stock. Care should be taken that all pages of a manuscript are in proper order, are numbered in sequence (with extra inserted pages clearly marked), and that copy for cuts is properly identified by annotations (since some printers do not make their own plates and have to send cut copy out to an engraver). All shipments of copy should be accompanied by a record list (duplicate retained) and should be made by first-class mail or Railway Express.

Modern publishing procedures involve "transcribing" the original copy by some method, usually Linotype or Monotype letterpress typesetting, to give a "shop proof" which is proofread against the copy by the proofroom, after which the type or transcription is corrected and a clean "galley proof" sent to the author (usually) or to the editor. This proof should be read against copy again, and all needed corrections should be marked. Authors should be reminded that many journals charge the cost of proof *changes* to them. They usually return corrected proofs to the editor, accompanied by an order for reprints. Since few authors are versed in proofreading and correcting, each author's proof should then be read by someone in the editorial office to approve alterations, check queried details, prevent expensive rewriting, and avoid style-breaking changes.

Most technical journals keep their articles in process "inventory" rather fluid, and plan each consecutive issue to contain such papers as have been promised and as many others as are available from authors. Editorial offices sometimes prepare dummies of each issue, but more often they supply the printer with an order of make-up list, including page running heads, and with a cover contents list.

The printer then corrects the type (or master copy for photo-reproductions), makes up the pages, and sends page proofs to the editor. These, too, should be read or checked, verifying all galley proof changes, and inspecting and approving or correcting many other details, such as running heads, page sequence, footnote placement and sequence, cross references, and locations of tables and cuts. It is not stretching the truth to remark that the work of the managing editor may make or break a journal's budget.

CONCLUSION

It probably is safe to say that every technical-journal editor, whether of a large or small journal, has entered upon his assignment with mental reservations and misgivings, a feeling of lack of training and competence for the job, and a certainty of short tenure. After learning the editorial ropes, becoming acquainted with his reviewers, and finding that contributors are human and understanding, too, the outlook brightens. Longer experience brings a feeling of real progress and achievement, and a subconscious realization that the editor has the privilege and duty of serving as a minister, physician, lawyer, and teacher to those scientists whose literary souls have been placed in his charge.

Chapter Nine

Technical Editing: The Reviewer's Viewpoint

LOUIS MEITES
Associate Professor of Analytical Chemistry
Polytechnic Institute of Brooklyn
Brooklyn, New York

It is said that there were once six wise men of Indostan, to learning much inclined, who went one day—though all of them were blind—to inspect an elephant in an attempt to discover its essential characteristic. The consequences of their expedition are, of course, too familiar to require description. Their ludicrous conclusions resulted, of course, from the fact that each of the blind men acquainted himself with only a portion of the subject of their investigation.

The same danger accompanies any effort to describe the viewpoint, function, and *modus operandi* of the reviewer of technical papers. Like the elephant, the reviewer has a good many essential characteristics, not all of which can be perceived from any one vantage point. This chapter consists, therefore, of an attempt to describe the five kinds of responsibilities which rest on the reviewer's shoulders, and the ways in which these different (and often conflicting) responsibilities manifest themselves in the preparation of a referee report.*

Let us begin by describing these five duties very briefly before proceeding to details. The first responsibility, of course, is to the journal's editor, who must decide whether the manuscript is or can be made into a sufficiently valuable contribution to the scientific literature to merit publication. The second is to the author, who, through a too prolonged and detailed concern with the raw material of his paper, may have fallen into errors of interpretation, calculation, or judgment which weaken the presentation of his results. The third is to the future reader of the published manuscript, who is entitled to the clearest (and most detailed) possible description of the author's accomplishments. The fourth is to the great majority of subscribers to the journal, who, though they may never read or even see the paper in question, must nevertheless pay for the space it consumes. The fifth is to the authors of the papers cited in (or omitted from) the list of references.

RESPONSIBILITY TO THE EDITOR

The reviewer of a technical paper is a scientist selected by the editor for his acquaintance with and competence in the field of the paper. As such, it is his responsibility to supply the editor with an informed judgment of the importance of the paper's subject, and also of the value of the contribution which the paper makes to the specialist's knowledge of that subject.

The criteria of acceptability naturally vary somewhat

* It amplifies the journal editor's views on the subject, expressed in the preceding chapter.

from one journal to another. For a journal of "pure" science, such as the *Journal of the American Chemical Society* or the *Journal of Physical Chemistry,* it would be difficult to improve on Professor W. A. Noyes, Jr.'s statement that an acceptable paper is one which either makes a substantial contribution to the body of scientific theory or contains numerical data which will stand the test of time. To these two categories the "applied" journals, such as *Analytical Chemistry,* add a third: that of papers which propose substantial improvements of procedure.

The reviewer's first reading of a manuscript serves primarily to allow him to form a judgment as to whether one of these criteria is met. The well-known pressure to publish seems to be continually increasing from one decade to the next. Both academic and industrial scientists are well aware that their professional reputations, their promotions, and in some deplorable cases even their very jobs, depend at least in part on their record of publications. As often as not, these records are scrutinized by administrators who are incapable of evaluating the merits of the papers' contents; under such circumstances, it is hardly to be wondered at if the quality of a paper becomes a consideration secondary to its mere existence in print and on a list of titles. In view of the rewards which may be his if he piles Pelion on Ossa in this fashion, the author eager for creature comforts may be driven to write and submit a paper based on scanty results hastily obtained; on a theory which accounts for some observations with which he has been immediately concerned but not others which he may have overlooked; on a minor modification of some familiar procedure which makes no real improvement or fills no real need; or even on material covered by an earlier paper which he has failed to consult.

When confronted by such a manuscript, the reviewer's task is a short and simple one. The editor must be informed that the manuscript is not comparable in quality to other work currently being published in the same field, and must be apprised of the reason why. Usually this takes only a few sentences, and is followed by rejection of the paper.

The reviewer's life would be a rather dismal one if manuscripts of this kind were very common. Fortunately, however, they actually constitute a fairly small minority of the papers submitted to reputable scientific journals. Not many authors are willing to succumb so far to the pressure to publish as to blemish their reputations in the eyes of their colleagues by publishing or attempting to publish papers with so little merit.

ASSISTANCE TO THE AUTHOR

Once he has satisfied himself that a manuscript contains something worth publishing, the reviewer should (unfortunately it is not possible to say "does") proceed to make himself intimately familiar with its contents. He should study the experimental procedures used, examine the data obtained, check the derivations of any equations newly derived, and carry out one or two typical calculations, all with a view to becoming as much the master of the paper's raw material as the author himself. He must think his own way through the author's assumptions and conclusions, and must examine all the consequences of these assumptions and conclusions.

This is, of course, the most time-consuming and demanding of the reviewer's duties. It is also the one most often neglected because of overwork, haste, and sometimes perhaps even incompetence. In my opinion, however, it is not only the most important of the reviewer's many functions: it is virtually the only real justification for the inclusion of the reviewer in the technical-editing process.

Ideally, the nature of scientific research is such that the author should be the world's leading authority on his own work and its relation to the general body of scientific knowledge and theory. There are, however, several reasons why this is often not the case. One is the well-known difficulty of seeing a forest after a long close examination of its trees. The reviewer, who has not been involved in the tortuous process of solving the problem, and who has therefore not been misled by incorrect hypotheses adopted during the research, is often better able than the author to discern substantial errors of fact or interpretation. Frequently the reviewer is engaged in research closely related to that described in the paper, and as a result has data at his disposal which are unknown to the author but which may be of material assistance in the interpretation of the author's own data. Finally, it is not at all uncommon for the reviewer to be a man of so much experience, knowledge, and eminence in the author's field that he is actually better able to interpret the data than the author himself.

For one or another of these reasons, the reviewer is usually in a position to render enormous assistance to the author by pointing out minor errors which the latter may have overlooked, proposing crucial experiments which would more conclusively bolster the interpretation, or suggesting a totally new interpretation which provides a better explanation, not only of the author's data, but also of other data of which the author may not have known. The reviewer who accepts the responsibilities of this position thereby makes a contribution to the scientific literature which is no less real than if he were exerting the same effort on a paper of his own. In the latter event, of course, the gain to himself is visible to all; in the former the gain is all on the side of the author, who usually knows not whence it comes. But to draw a distinction between the two cases on such a basis is to substitute selfishness for science.

I would like to dwell on the phrase "knows not whence it comes" for a moment, for this touches on what has certainly been one of the most bitterly disputed features of the process of technical editing: the anonymity of the reviewer. Surely this has never been a matter of any concern when the critical scrutiny just described has led the reviewer to conclude that the paper deserves publication with no more than relatively minor modification. However, it is not at all incomprehensible that the author who has spent months or years gathering data and developing what seems to him to be a perfectly reasonable explanation of the phenomena he has observed should be both chagrined and resentful when the entire structure is abruptly demolished by an unknown referee.

Let us be perfectly clear about two facts. One is that referees are by no means possessed of an inhuman detachment which prevents them from wincing when authors tread on the toes of their favorite beliefs. Nor does a scientist suddenly don a mantle of infallibility on receiving an editor's invitation to pass judgment on a paper. Reviewers, like judges, are sometimes biased, careless, and even downright wrong.

The institution of reviewer anonymity is founded on several premises which do not always correspond exactly to the truth. One is that the reviewer is completely free from personal or scientific bias of any kind. Another is that he is sufficiently interested in ensuring that the paper finally published is of the highest possible quality to be willing, when the need arises, to devote a good many hours to a task from which he personally stands to gain nothing whatever. A third is that he is sufficiently well versed in the field to be able as well as willing to lend substantial help to the author. If these conditions are fulfilled, then the reviewer serves merely as a representative of those who would scrutinize the paper when it appeared in print, and as such his own personality and identity are irrelevant. If they are not fulfilled, there is no recourse but for the editor to solicit the advice of other reviewers who are either more impartial, more conscientious, or better informed.

I mentioned that reviewers are by no means infallible in their disagreements with authors, though I would add that in my opinion I have more often been right in such a disagreement when I was a reviewer than when I was an author. In any case, the author is not deprived of an opportunity to reply to even the most far-reaching scientific criticism. The reviewer may have overlooked an important link in the author's chain of evidence or reasoning, or he may have been misled by a carelessly-phrased portion of the manuscript. If the author's considered judgment is that the reviewer has erred, it is his duty to search for the cause of this error and, in company with reviewer and editor, to try to find a way of modifying the paper so that others will not be similarly misled by the version finally published. In all such cases, of course, truth—like murder—will out.

RESPONSIBILITY TO READERS

When the reviewer has persuaded himself of the essential validity of the author's work and conclusions, and has decided that publication of the paper will advance rather than harm the author's scientific reputation, his responsibility to the author ends for a time, and he should place himself in the position of the reader of the paper. This may involve putting the paper aside for a few days to allow the trees to recede into the forest, then taking it out again for still another reading.

In this step the reviewer's concern is with the manner of presentation. Authors often submit what can only be described as histories of their research: that is, manuscripts in which the material is arranged more or less chronologically. When this happens, the reviewer should try to ease the reader's task by suggesting a more logical structure, one in which the successive ideas and confirmatory experiments are arranged in the manner of a compound syllogism, so that the reasons for and meanings of the successive steps become plain. An important new thought may have been dismissed in a subordinate clause, while a dead horse may have been flogged up and down two or three pages. Hypothesis and experimental test may have been separated, while a large number of data may have been so confusingly thrown together that their connection and significance are hard to perceive. Figures are sometimes so poorly drawn, and tables so thoughtlessly arranged, that they can be interpreted only with great difficulty. Sometimes an ambiguous or unclear sentence or paragraph may be the cause of great confusion. Many highly competent scientists are but indifferent technical writers, though it is doubtful whether most are as incompetent in the use of the English language as certain editorial-office policies would seem to suggest. Since the reviewer has already gone to the trouble of finding out exactly what the author is saying—or, more accurately, what the author is justified in saying—it costs him very little more effort to make certain, insofar as it lies within his power to do so, that the author's message is expressed in the clearest and most lucid possible fashion.

Both by some reviewers and by some copy editors or editorial assistants, this process is occasionally perverted into a schoolma'amish red-penciling of the author's grammar, syntax, and punctuation. It is impossible to register too strong a protest against this kind of subversion of a legitimate striving for clarity and editorial uniformity, especially when it is committed by nontechnical people, who usually manage to butcher an author's meaning while homogenizing his grammar. Certainly it behooves the reviewer to tread softly in such matters. An absolute barbarism or a typographical error is usually something to which the author appreciates having his attention called, but with this the reviewer's responsibility ends. A referee report which is primarily devoted to matters of grammar and punctuation tends to make both editor and author wonder whether the reviewer has bothered to notice the scientific content of the paper.

RESPONSIBILITY TO THE GENERAL READER

Though a complete and detailed description of the author's work and conclusions is essential in every paper, prolixity costs money and wastes the time of readers. It is quite natural to believe that one's own paper is the most important of all those which appear in any issue of a journal, and to provide detailed descriptions of our life-and-death struggles with incidental problems. But space is money, and the reviewer must often search for material that can be deleted without seriously affecting the force of the author's conclusions. Figures can occasionally be excised wholesale; large masses of tabular data can be made available to the specialist via a microcopy and photoprint service; and the incidental problems which were so troublesome to the author can often be relegated to a position which more accurately reflects their importance in the over-all structure.

RESPONSIBILITY TO SUBJECT AUTHORITIES

The reviewer's final duty is to the authors who have blazed the trail along which the paper being considered represents the latest advance. Of course it is not at all uncommon for the reviewer himself to be numbered among these. This is both a rather simple process, and, for the reviewer who is as well acquainted with the literature as he should be, a quite automatic one. It consists of making certain that no important relevant paper has been overlooked, and also that the author has properly understood and described the significance of each of the principal papers he has cited. Naturally, the reviewer is in no position to conduct an extensive literature survey. But he may know of a significant paper on a related subject which the author has overlooked, or he may feel that the author has misinterpreted the work of some earlier scientist. In either case it is his responsibility to call the oversight or error to the author's attention.

THE REVIEWER'S REPORT

Having completed these five essential steps, the reviewer turns to the preparation of his report. The length and complexity of this document naturally varies tremendously from one paper to another. Once in a while it consists merely of a terse recommendation that the manuscript be accepted as it stands; about as rarely it consists of an equally brief recommendation that the paper be rejected.

It is ordinarily unnecessary to list one's reasons for recommending the acceptance of a paper. It is true, however, that many such recommendations are written for papers which are perfectly routine in every way. For this reason it would seem to be no more than simple courtesy to add a brief explanation to the recommendation in dealing with a manuscript of unusual excellence. An unqualified recommendation for rejection is much more difficult to write, for it is necessary to explain the shortcomings of the manuscript to the editor, who is rarely an expert in the field of the manuscript.

Probably no less than four-fifths of referee reports fall into neither of these classes, but are qualified recommendations of acceptance subject to the author's consideration of the points which have come up during the reviewing process. Only one specific rule for the preparation of such a document need be suggested. This is that no criticism should be made without both a clear explanation of the reason for the criticism and a brief suggestion of how the criticism could be met. The primary purpose of a report of this type is to help the author to prepare a better manuscript, and there is no room here for destructive criticism.

With the submission of this report, the reviewer's connection with a paper ordinarily comes to an end. Sometimes, however, this is not the case, and he is then confronted with a revised manuscript and a copy of a letter of explanation from the author. This usually results from the author's disagreeing with the reviewer's comments, coupled with an editor's reluctance to make a decision in the face of what appears to be a dispute of substantial proportions.

At this point the entire process must commence again, with the reviewer now being particularly scrupulous in reexamining all of the premises and thought processes which led him to his earlier conclusions. If this re-evaluation leads him to the conclusion that the author has substantial grounds for the statements in question, the paper should generally be allowed to pass without further ado. If, on the other hand (as sometimes happens), the author's revision and reply merely reiterate without lending additional support to his earlier position, the reviewer is probably best advised to prepare a concise summary of his objections and request that the entire matter be passed along to a member of the journal's editorial board for final arbitration.

The process of technical editing is one in which author,

reviewer, and editor must cooperate wholeheartedly toward their common aim of ensuring the highest possible quality of the final product. The competence, integrity, and selflessness of the reviewer are essential to the achievement of this aim and to the preservation and strengthening of the standards we have all come to expect of our technical journals.

Chapter Ten

Editing the Technical Business Magazine

WILLIAM F. BLAND*
Editor, Petroleum Processing
New York, New York

Perhaps you visualize the editor of a technical business magazine as a man who is concerned mostly with gathering information, putting it together, and then getting it down on paper for the benefit of some reader.

Or maybe you think of him as sitting at a desk, wielding a blue pencil and pair of shears, and cutting your article to pieces.

I'd like to give you a somewhat different concept of a technical-business-magazine editor. He's what I call a "communication expert." He may at times work at reporting and writing, and again at editing another person's copy. But his primary job is communication. He must get the ideas that are in any good article *off the printed page* and *into the reader's mind* as clearly and quickly as possible.

* Now Assistant to the Publisher, Petroleum Publications, McGraw-Hill Publishing Company, New York, New York.

Let me re-emphasize that point. His job does not end with getting an article into print. It ends only with getting it *off the printed page* and *into the reader's mind.*

I should make it clear that I am not talking only about the top or chief editor on a magazine. These ideas apply to any person on the editorial staff who is responsible for producing or handling copy.

WHAT WE MEAN BY COMMUNICATION

When I'm explaining this concept of what a technical business magazine editor does to a newcomer on the staff, I do it something like this: Let's assume we have an article on the subject of radiation which we're getting ready for our magazine.

Now, no single article, of course, will interest all our readers. But for the sake of a starting point, let's assume that 10,000 of them could get something of value out of reading this particular article. To do a good job of technical editing, we should get as many as possible of those 10,000 readers to read it.

Some of them we don't have to worry about. They're the "eager beavers." They will avidly read everything and anything they find on radiation, even if it is set in type of the size that you find in classified advertising, and run solid.

We can get others to read the article with only a little effort on our part. They are the ones who leaf through a magazine; if you can catch their attention and get them to stop, they'll usually start reading. We can put an illustration of some sort at the top, to catch the eye of this sort of reader as he is scanning the magazine, and instead of using a label we can get a little more action into the headline—"Going into Radiation?".

But there are other potential readers who are more diffi-

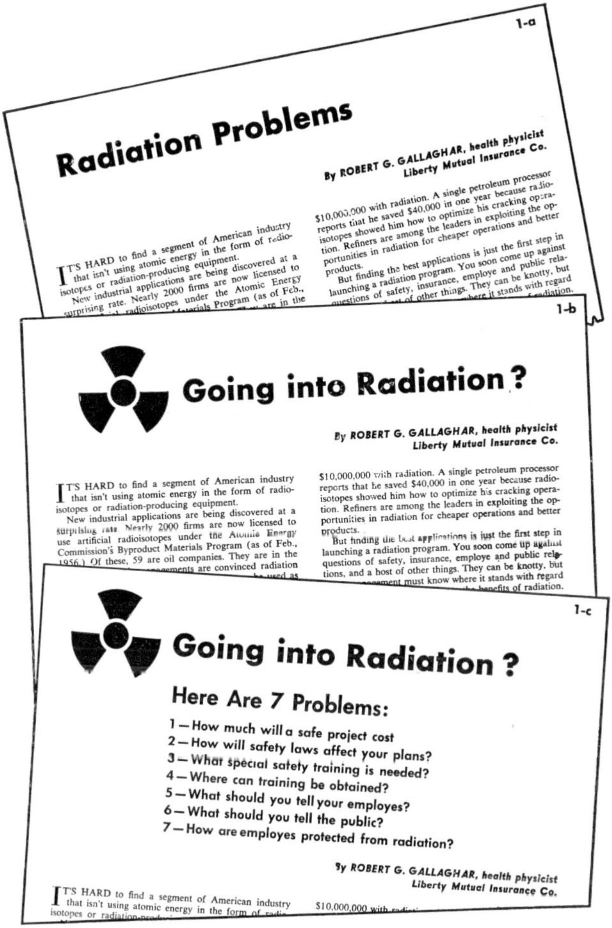

Figure 10.1. Communication starts right at the top of an article—with the opening illustration and the headline. Part 1-a is the normal "dry" way to start a technical article, with a label heading and no illustration. Part 1-b adds an illustration and peps up the headline. Part 1-c gives a capsule abstract in the form of a secondary heading which tells at a glance what the article's main points are.

cult to catch. Perhaps they are busy, and are going through the magazine rapidly. Or perhaps they have other things on their minds, or are a little impatient. They have to be spoon-fed, so to speak. To stop them—to make readers out of them—we must spell things out a little more clearly. We must present key bits of information in such a manner that the scanner can at a glance pick up the main ideas of what the article is about. We have to sell him on why he should stop and read. We'll have a better chance of getting these readers if our headline says something like: "Going into Radiation? Here are 7 Problems," and then list the seven problems immediately under the headline. Figure 10.1 shows the three ways in which this article could be presented.

Please don't get me wrong. We *do not* mislead the reader. We *do not* promise him something which isn't in the article. We *do* try, by whatever techniques we can, to get him to start reading. The whole point is that an article is no good if it isn't read, regardless of how much helpful information it contains and how well it is written.

You may be asking, "Why bother with these 'marginal' readers? If they're in such a hurry, or too lazy to dig out the meat of an article for themselves, that's their tough luck." I contend that it is the editor's job to convince the marginal reader that he should read the complete article, just as it's the highway patrolman's job to convince the reckless driver to slow down and save his neck. Granted, the policeman can get a little tougher than the editor—but neither of them can afford to shrug his shoulders and forget his responsibilities.

There's another point. The technical man today has a lot to read, plus heavy demands on his time from other quarters. The technical-business-magazine editor is doing a real job when he can make it easier and faster for his reader to be more decisive and correct in his choice of what he reads.

HOW DO WE ACHIEVE COMMUNICATION?

The secret of successful communication in the technical business magazine is "presentation." And by "presentation" I mean the manner in which an article is presented to the reader—what the headline says, the illustrations that are used and how they are used, and how the article is organized. Proper presentation gets an article moving fast—gets it off the printed page and into the reader's mind—as quickly as possible, and with the least effort on the reader's part. Good presentation makes an article "pop," so that the reader who is merely scanning a page can pick up the main point at a glance. We fully expect, of course, that he'll stop to read the whole piece if he's interested in the subject, and if we've done our part properly.

There are many ways in which presentation of a technical article can be improved, but I'll confine myself here to five of the major ones: (1) organize and structure the article, (2) select the proper illustrations, (3) make the illustrations work, (4) use optional reading, and (5) use color and typography.

Organize and Structure the Article

Every well-written technical article, of course, starts with a good outline of the topics the writer is going to cover. All too often, however, the outline gets lost in the article—even though it's still there. Good editing will bring the outline into the open, and good presentation will make it readily apparent to the reader. This can be achieved with the proper use of type, such as large sectional headings throughout the article, smaller but still prominent subheads, and bold-faced "lead in" words or phrases at the start of key paragraphs.

These serve as road signs or guide posts for the reader, to lead him through the article and show him quickly how it is organized and where he is. Figure 10.2 gives some examples of how the organization or structure of an article can be made readily apparent.

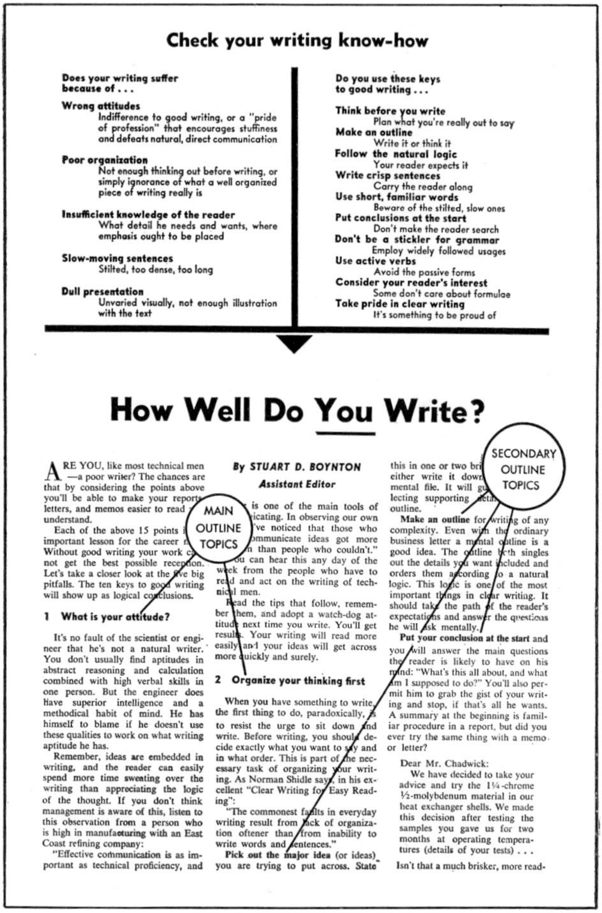

Check your writing know-how

Does your writing suffer because of . . .

Wrong attitudes
Indifference to good writing, or a "pride of profession" that encourages stuffiness and defeats natural, direct communication

Poor organization
Not enough thinking out before writing, or simply ignorance of what a well organized piece of writing really is

Insufficient knowledge of the reader
What detail he needs and wants, where emphasis ought to be placed

Slow-moving sentences
Stilted, too dense, too long

Dull presentation
Unvaried visually, not enough illustration with the text

Do you use these keys to good writing . . .

Think before you write
Plan what you're really out to say
Make an outline
Write it or think it
Follow the natural logic
Your reader expects it
Write crisp sentences
Carry the reader along
Use short, familiar words
Beware of the stilted, slow ones
Put conclusions at the start
Don't make the reader search
Don't be a stickler for grammar
Employ widely followed usages
Use active verbs
Avoid the passive forms
Consider your reader's interest
Some don't care about formulae
Take pride in clear writing
It's something to be proud of

How Well Do You Write?

ARE YOU, like most technical men—a poor writer? The chances are that by considering the points above you'll be able to make your reports, letters, and memos easier to read ... understand.

Each of the above 15 points i... important lesson for the career ... Without good writing your work ca... not get the best possible reception. Let's take a closer look at the five big pitfalls. The ten keys to good writing will show up as logical conclusions.

1 What is your attitude?

It's no fault of the scientist or engineer that he's not a natural writer. You don't usually find aptitudes in abstract reasoning and calculation combined with high verbal skills in one person. But the engineer does have superior intelligence and a methodical habit of mind. He has himself to blame if he doesn't use these qualities to work on what writing aptitude he has.

Remember, ideas are embedded in writing, and the reader can easily spend more time sweating over the writing than appreciating the logic of the thought. If you don't think management is aware of this, listen to this observation from a person who is high in manufacturing with an East Coast refining company:

"Effective communication is as important as technical proficiency, and

By STUART D. BOYNTON
Assistant Editor

... is one of the main tools of ...icating. In observing our own ...'ve noticed that those who ...ommunicate ideas got more ...n than people who couldn't." ...ou can hear this any day of the week from the people who have to read and act on the writing of technical men.

Read the tips that follow, remember them, and adopt a watch-dog attitude next time you write. You'll get results. Your writing will read more easily and your ideas will get across more quickly and surely.

2 Organize your thinking first

When you have something to write, the first thing to do, paradoxically, is to resist the urge to sit down and write. Before writing, you should decide exactly what you want to say and in what order. This is part of the necessary task of organizing your writing. As Norman Shidle says, in his excellent "Clear Writing for Easy Reading":

"The commonest faults in everyday writing result from lack of organization oftener than from inability to write words and sentences."

Pick out the major idea (or ideas) you are trying to put across. State this in one or two bri... either write it down... mental file. It will gu... lecting supporting deta... outline.

Make an outline for writing of any complexity. Even with the ordinary business letter a mental outline is a good idea. The outline both singles out the details you want included and orders them according to a natural logic. This logic is one of the most important things in clear writing. It should take the path of the reader's expectations and answer the questions he will ask mentally.

Put your conclusion at the start and you will answer the main questions the reader is likely to have on his mind: "What's this all about, and what am I supposed to do?" You'll also permit him to grab the gist of your writing and stop, if that's all he wants. A summary at the beginning is familiar procedure in a report, but did you ever try the same thing with a memo or letter?

Dear Mr. Chadwick:
We have decided to take your advice and try the 1¼-chrome ½-molybdenum material in our heat exchanger shells. We made this decision after testing the samples you gave us for two months at operating temperatures (details of your tests) . . .

Isn't that a much brisker, more read-

Figure 10.2. The organization of an article can be made readily apparent by the proper use of typography. Bold-face headings and words serve as guide posts to lead the reader through the article.

Select the Proper Illustrations

Good illustrations can set the stage for an article, can get a key idea over to a reader quickly. They "show" what it is sometimes difficult to "tell." An illustration can be almost anything pictorial—photograph, flow chart, drawing, graph, map. If we have an article without illustrations, we'll try to make some. Cartoons are sometimes good, and even a tabulation of a few key facts can be used, as shown in Figure 10.3. But to be good, the illustration must tell or explain something to the reader. It's a mistake to concoct an illustration simply for the sake of having it.

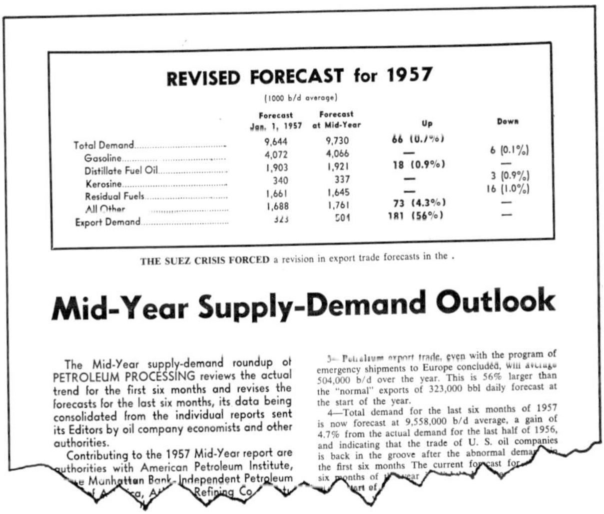

REVISED FORECAST for 1957

(1000 b/d average)

	Forecast Jan. 1, 1957	Forecast at Mid-Year	Up	Down
Total Demand	9,644	9,730	66 (0.7%)	
Gasoline	4,072	4,066	—	6 (0.1%)
Distillate Fuel Oil	1,903	1,921	18 (0.9%)	—
Kerosine	340	337	—	3 (0.9%)
Residual Fuels	1,661	1,645	—	16 (1.0%)
All Other	1,688	1,761	73 (4.3%)	—
Export Demand	323	501	181 (56%)	—

THE SUEZ CRISIS FORCED a revision in export trade forecasts in the .

Mid-Year Supply-Demand Outlook

The Mid-Year supply-demand roundup of PETROLEUM PROCESSING reviews the actual trend for the first six months and revises the forecasts for the last six months, its data being consolidated from the individual reports sent its Editors by oil company economists and other authorities.

Contributing to the 1957 Mid-Year report are authorities with American Petroleum Institute, ... Manhattan Bank, Independent Petroleum ... Refining Co.

3—Petroleum export trade, even with the program of emergency shipments to Europe concluded, will average 504,000 b/d over the year. This is 56% larger than the "normal" exports of 323,000 bbl daily forecast at the start of the year.

4—Total demand for the last six months of 1957 is now forecast at 9,558,000 b/d average, a gain of 4.7% from the actual demand for the last half of 1956, and indicating that the trade of U. S. oil companies is back in the groove after the abnormal dema... the first six months The current forcast for ... six months of ... year ... start of ...

Figure 10.3. Tabulations make good illustrations if they don't include too much data, are kept simple, and are set in heavier and bolder type.

Make the Illustrations Work

We'll frequently use an illustration for two purposes: first, to attract the attention of the person who is casually leafing through the magazine; secondly, to show something more clearly than mere words can describe it.

In the first case, we'll look for some key illustration that tells what the article is about. When such an illustration is properly placed (at the top of the article, and above or to the left of the heading), it serves as an "eye magnet." It

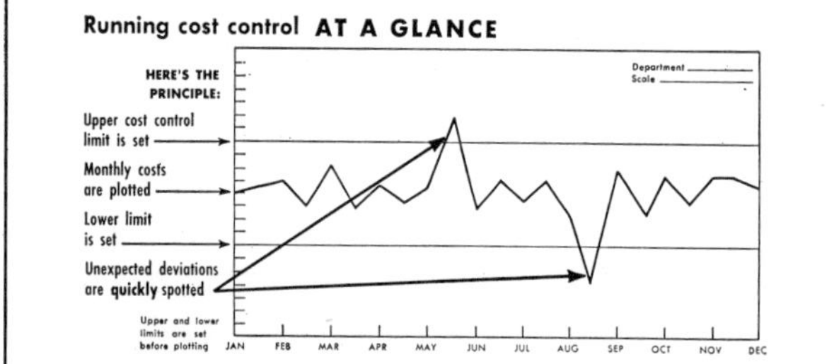

'Instrument Charts' Track Costs

By J. N. WATSON
British American Oil Co.

It's new and important. British American in Canada has developed a control system that applies, for the first time, particularly efficient statistical methods to refinery cost control. Results to date are encouraging

HIGHLY DEVELOPED process control, based on instrumentation, is permitting small teams of operators to observe and control large, complex processes. A similar technique is being develop...

Why We Tackled The Problem

We recognized that severe limitations were imposed on interpreting and controlling expenses by the classical ...

unit of throughput" in expressing expense has been common practice in the petroleum industry for many years, the example in Fig. 1 graphically shows ...

Figure 10.4. Illustrations that work with the headline in an article not only serve as "eye magnets" to catch the reader's attention, but also help tell him what the story is about.

catches the reader's attention, and then his normal eye movement leads him into the heading. If the illustration and the heading say the same thing, that's good. Our radiation symbol in Figure 10.1 is a good example, and Figure 10.4 is another.

In the second case, we'll look for ways to make the illustrations more effective. Maybe we can do this by running a photograph of a piece of equipment alongside a drawing of it—preferably in the same scale and from the same angle. Sometimes we can simplify a drawing or flow chart by eliminating some of the lines or words. We'll place "before" and "after" pictures side by side, to give the reader a faster comparison.

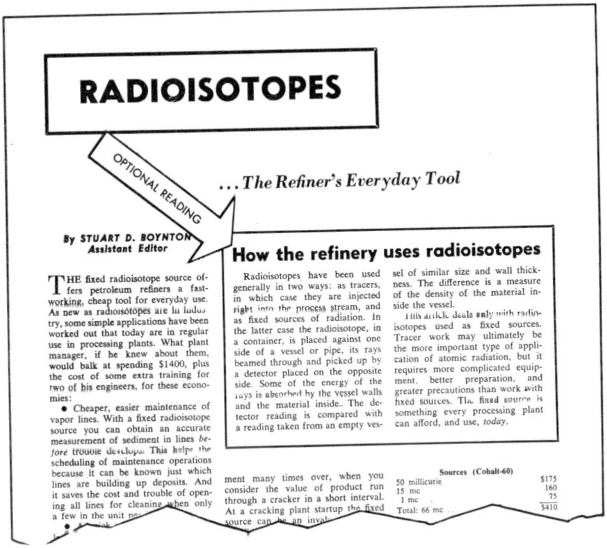

RADIOISOTOPES

OPTIONAL READING

...The Refiner's Everyday Tool

By STUART D. BOYNTON
Assistant Editor

THE fixed radioisotope source offers petroleum refiners a fast-working, cheap tool for everyday use. As new as radioisotopes are in industry, some simple applications have been worked out that today are in regular use in processing plants. What plant manager, if he knew about them, would balk at spending $1400, plus the cost of some extra training for two of his engineers, for these economies:

- Cheaper, easier maintenance of vapor lines. With a fixed radioisotope source you can obtain an accurate measurement of sediment in lines *before* trouble develops. This helps the scheduling of maintenance operations because it can be known just which lines are building up deposits. And it saves the cost and trouble of opening all lines for cleaning when only a few in the unit

How the refinery uses radioisotopes

Radioisotopes have been used generally in two ways: as tracers, in which case they are injected right into the process stream, and as fixed sources of radiation. In the latter case the radioisotope, in a container, is placed against one side of a vessel or pipe, its rays beamed through and picked up by a detector placed on the opposite side. Some of the energy of the rays is absorbed by the vessel walls and the material inside. The detector reading is compared with a reading taken from an empty vessel of similar size and wall thickness. The difference is a measure of the density of the material inside the vessel.

This article deals only with radioisotopes used as fixed sources. Tracer work may ultimately be the more important type of application of atomic radiation, but it requires more complicated equipment, better preparation, and greater precautions than work with fixed sources. The fixed source is something every processing plant can afford, and use, *today*.

ment many times over, when you consider the value of product run through a cracker in a short interval. At a cracking plant startup the fixed source can

Sources (Cobalt-60)	
50 millicurie	$175
15 mc	160
1 mc	75
Total: 66 mc	$410

Figure 10.5. Optional reading material can be put to one side where it's available for the reader who wants it—but where it doesn't slow up the main story.

Use "Optional Reading"

Very often there will be some bit of information in a story which is interesting and helpful to some readers but not essential. This slows up the article if left in, but if it is removed we don't have a complete story for some readers. The solution here is to remove the material in question from the main text and run it as "optional reading"—as a box or a separate supplementary chunk of reading matter, as shown in Figure 10.5. The reader who doesn't want this information can bypass it easily; the reader who does want it has it available.

Use Color and Typography

We try to use color only when there's some good reason for it—when it can do a job such as highlighting a particular line or area in a chart or drawing, or calling attention to some key words.

TEL Makers Expand to Match Market

The two manufacturers of octane-increasing tetraethyl lead are now jockeying for position to supply what they are convinced will remain a steadily-growing market.

TEL, generally conceded to be the gasoline additive with the best anti-knock characteristics, is made principally from pig lead, sodium, and petroleum derivatives.

Up until 1948, the compound was manufactured only by E. I. du Pont de Nemours & Co., and marketed by the Ethyl Corp. In 1948, however, Ethyl entered into its own TEL manufacturing, and Du Pont set up its own TEL marketing organization.

Now both companies are moving to keep ahead of future demand.

This week, for instance, Du Pont is completing conversion of a dry-cargo-vessel-turned-tanker (picture) which will facilitate its TEL shipments along the Gulf and East Coasts (PW—Jun.15'56,p47).

The "new" tanker—at 6,000 tons deadweight—has been leased from Valentine Chemical Carriers, Inc. Beginning next month, it will be the first vessel to carry tanker-quantity shipments of TEL. It will move TEL from Du Pont's Chambers Works production plant at Deepwater, N. J., to the company's Beaumont (Tex.) TEL distribution center.

On return trips, the tanker will pick up TEL raw materials—ethylene dichloride at Port Neches, Tex., and ethyl chloride at Houston—for delivery to the Chambers works.

Du Pont only recently put a second TEL plant—at Antioch, Calif.—into production (picture).

The new multimillion-dollar plant will manufacture TEL and ship it to refiners west of the Rockies.

The Antioch plant—on the San Joaquin River up from San Francisco Bay—uses a new, continuous process for making TEL, said to be more efficient than the older batch process. The plant has all manufacturing equipment out of doors rather than inside buildings, to facilitate expansion.

Ethyl Corp., too, is building a new West Coast plant, its fourth in North America (PW—Feb.22'57,p75).

Ethyl's new plant will be at Pittsburg, Calif., just a few miles from Antioch. Construction has just begun, with completion scheduled for mid-1958.

Ethyl has other TEL manufacturing plants at Houston, Baton Rouge, and Sarnia, Ont. And it has property in Joliet, Ill., for another plant site.

Ethyl is also getting ready to announce a new process for making TEL. The company is expected to make known the new process some time next month.

Back of all these advances is the increasing demand, not just for gasoline, but for high-octane gasoline.

TEL manufacturers take as much delight in the increasing octane requirements of newer cars as they do in the rapidly growing car population. With no end in sight for the current trend to higher-compression-ratio engines, gasoline octane ratings are expected to keep right on rising.

At present, finished gasoline octane ratings can be boosted substantially by adding TEL. So Du Pont and Ethyl are convinced that refiners will lean more heavily toward the TEL additive in the future.

Figure 10.6. Variable-speed reading as practiced by *Petroleum Week* enables the busy reader to get the highlights by reading only the bold type, or to pick up details by reading the light type also.

Plain, simple type can also be used to advantage, as has already been discussed briefly in connection with the use of secondary and sectional headings within a story to show how it is organized.

One interesting use of typography has been developed during the past few years by my sister publication at

McGraw-Hill, *Petroleum Week*. It involves printing the main facts in a story in bold, heavy type, and the supporting details in regular light type. The result, an example of which is shown in Figure 10.6, is an article designed for "variable-speed reading," by means of which a reader can quickly get the highlights in skeleton form by reading only the heavy, bold type throughout the article, or can pick up the details on any particular point by reading the light type.

Obviously, this technique requires some very special editorial skill and training. The editor has to write a story within a story, making the skeleton story read smoothly without being jumpy, and at the same time making it fit into the complete story.

I'm not sure I'd recommend this technique for an involved technical article, but it might be interesting to experiment with.

THE MEETING PAPER, THE SOCIETY JOURNAL, AND THE TECHNICAL BUSINESS MAGAZINE

The editorial philosophy and the techniques I've been discussing are not, of course, peculiar to the technical business magazine. To some degree they can be applied with advantage to any form of written communication, and to some forms of verbal communication.

In the area of publications which this book covers, however, I think this philosophy and these techniques are particularly important for the technical-business-magazine editor. He has the best opportunities and the greatest need for applying them.

The author of a technical-meeting paper doesn't have very much control over the audience he will have. The people who will attend the session, to listen to him, make up their minds to do so in advance, and pretty much on the basis of their interest in his topic. Once the author starts presenting his paper, individuals in the audience are usually too polite to get up and walk out, regardless of how poor a job he is doing.

The editor of the technical-society journal is somewhat restricted as to how much he can do in the way of revising, reorganizing, and rewriting the contents of his magazine. Except in the larger societies, such as the American Chemical Society, he usually presents his society's meeting papers pretty much as they were prepared. Some society journals, however, have been making good progress recently in efforts to get out of this rut.

The technical-business-magazine editor, on the other hand, has a pretty free hand in selecting the scope of his editorial subjects and in organizing his editorial material as he sees fit. And he needs this free hand. He has an audience which is rather fluid and transitory. He has to stop his readers in the first place and then keep them interested. As you well know, it's very easy to turn to the next page in a magazine.

His success and his publication's success depend on how good a job he does—on how much "reader acceptance" he creates.

This need of "reader acceptance" means that the technical-business-magazine editor has to be much more than a historian for his industry and a reporter of what has been happening. He has to be aware not only of where his industry is today, but also of where it will be tomorrow—and how it will get there. He has to be alert not only to the problems it faces today, but also to those it will likely have to solve tomorrow—and how it will solve them. And he should have this insight into the future even before many people in his industry have it.

These are requirements not usually placed on the meeting-paper author or on the society-journal editor (except in a very few societies). On the other hand, the technical-business-magazine editor has tremendous resources at his disposal. He travels a lot, he calls on many people in his industry, he attends numerous technical gatherings. He usually has free access to the top technical and management brains in his industry, and frequently enjoys their confidence. Even though he can't print much of the advance information he receives, he can add together the little pieces he gathers from here and there and can form a pattern of how events are shaping up.

SOURCES OF EDITORIAL MATERIAL

One question frequently put to me is: "Where do you get your editorial material?" We're speaking here, of course, of feature articles, the kind that are at least a couple of pages long. We're not talking about the short items, such as announcements of personnel changes, new equipment, and the like.

In general, you'll find that all the articles in a technical business magazine come from one or more of three general sources—are staff-written articles, contributed articles, or meeting papers. Individual magazines, depending on their staffing and editorial policies, will usually favor one source over another.

Staff-Written Articles

Staff-written articles, of course, are written by the magazine's own editorial staff. They might be of the "round-up" type, in which one of the editors pulls together, usually by personal interviews and correspondence, a composite picture of a particular situation of current interest, such as "Where Are We Headed with Premium Gasolines?" Or they might be of the "reporting" type, in which one of the editors will visit a plant or laboratory (after making suitable arrangements), gather the necessary data, and then write his article, such as "How Esso Cut Flaring by 85%."

Contributed Articles

Contributed articles are written by people who are not on the publication's own staff. They are usually written on request or assignment. In these cases, the editor has an idea for a story and approaches some authority in the field with the request that he write it, or he may ask the company concerned with the story to assign one of its technical staff to write the article for his publication.

Another source of the contributed article is the free-lance technical writer. These writers are usually steadily employed in industry and do their writing on the side, with the hope of augmenting their earnings. They are omnivorous readers, and often build up voluminous data files on subjects of interest to them but not necessarily connected with their day-to-day work. Once a free-lancer establishes a reputation in a field, and assuming that he is a good writer and chooses the proper subjects, he seldom has any trouble selling his work.

You can see, of course, that this term "contributed" is not entirely correct. Contributed usually implies "gratis" or "free," but many of the "contributed" articles in a technical business magazine are paid for, on a basis that I'll explain later.

Meeting Papers

A number of technical business magazines operate with relatively small editorial staffs, depending to a large extent on the papers presented at society and association meetings for their editorial content. Some technical societies require that papers presented at their meetings become the property of the society, and reserve first publication rights to their own journals. Others, however, particularly industry associations such as the American Petroleum Institute, Western Petroleum Refiners Association, and the like, make their papers freely available to any publication which wants them, requesting only a credit line.

There's a lot of good meat in many of these papers, but it usually takes adroit and laborious editing to make it stand out. Too often, unfortunately, a technical business magazine that uses meeting papers simply throws a paper into type and publishes it exactly as it was originally presented—which probably was without much regard for how it would look in print.

EDITORIAL PRACTICES

You might be interested in some of the usual practices followed by a technical business magazine. Each publication has its own policies, of course, but these are ours on *Petroleum Processing*, and I believe they are fairly typical.

Approval of Articles

Unlike most newspaper reporters, the technical-business-magazine editor is usually glad to submit his staff-written article back to the source for checking when the information for the article has been gathered during a personal tour of an individual plant or in discussion with company personnel.

There are two reasons why we always like to have a technical article which we write checked over. First, we've found that engineers and plant operating people talk much more freely and give us more information when they know they'll have a chance to see the written result and to correct any misunderstandings. Second, it's too easy to make a mistake in reporting highly technical details, and we like to know that what we're printing is correct.

When we do refer an article back to the source, we

always make it clear that it is being submitted only for checking of technical accuracy and detail, and to prevent the unintended disclosure of information which should not be published. We reserve the right to follow our own writing style. Whenever we receive a contributed article, we always make certain that the writer has secured the necessary approval from his company—if he is writing about a subject that concerns his employer.

One type of article we usually do not submit for approval is the staff roundup, where we pull together the results of interviews or correspondence with half a dozen or more people. Nor do we submit the articles we write that are based on meeting papers or company press releases.

Checking Proofs

We generally do not send galley or page proofs to an author for checking prior to publication, unless the article involves some very complicated formulas and the like. In those cases we usually ask the author if he will check proofs.

In other cases, if the author requests proofs we'll accommodate him if time permits. But we try to discourage the practice. It's too easy for an author to change his mind when he sees his written words in type and decide to say something slightly differently. What he doesn't realize—unless he knows something about printing—is that removing or adding a single word at the start of a paragraph frequently makes it necessary to set new type for the rest of the paragraph. Changing a single word on a page proof can cost as much as $10.

Honorariums and Reprints

As pointed out earlier, contributed articles aren't really contributed. Most technical business magazines pay a modest honorarium for material written by an outside author.

In our case, we pay for a manuscript when it meets both of the following conditions: (1) when it is submitted for publication in *Petroleum Processing* on an exclusive basis, and (2) when the content of the article does not promote or publicize a service or product which the author or his company has for sale. If either condition is not met, we put the literal interpretation on the word "contributed."

No outside author will ever get very rich writing technical articles. But if he's willing to place some value on the prestige attached to a byline, and on the pleasure of seeing his work in print, then he can persuade himself that he has a profitable hobby.

Our usual rate of payment for editorial manuscripts is $25 a page, based on the number of pages the material takes in the magazine. Space is figured to include all text, tables, formulas, notes, and illustrations supplied by the author.

Sometimes we are willing to go somewhat above that rate to obtain a manuscript of exceptional quality and from a recognized authority. We also have paid less.

Some companies have a policy which prohibits their employees from accepting payment for material which the employee writes as a company assignment. In these cases we usually suggest that we'll be glad to provide reprints in lieu of cash payment. Those are the only instances where we provide gratis reprints to an author.

CONCLUSION

I have not attempted, in the preceding discussion, to give a complete picture of how the technical business magazine is edited. Nor have I outlined all the aspects of how better presentation can make for better communication.

To cover both subjects adequately would require a whole book, and I hope that some day such a book will be written.

I do hope that I've gotten across my one main idea—that editing the technical business magazine is a job of communication, of getting ideas not only *onto* the printed page, but also *off* the printed page and *into* the reader's mind.

Any editor who is successful in achieving that goal is doing a good job.

Chapter Eleven

Editing Technical News Releases

GORDON H. BIXLER

Assistant Managing Editor, Chemical and Engineering News
Washington, D. C.

Every day, hundreds of technical news releases flood into the offices of newspapers and magazines throughout the country. They may be very complex, like a set that came out not too long ago on overthrowing the theory of parity. Or they may be very simple, like news of meetings or of changes in personnel. And, of course, they span the whole range between these extremes.

Problems facing an editor of a technical news publication are, first, which of these releases should he use; secondly, how should he use them—should he rewrite them completely, partly, or not at all; and thirdly, if he decides to rewrite, how should he go about doing so?

This chapter will not deal with how an editor selects releases to publish. After all, each magazine is different, and each editor selects releases for the audience he wishes to reach. Only experience tells him how to do this. Instead, the chapter will discuss how an editor handles releases once he has made his selection.

Actually, most news releases need some rewriting. Companies write releases for the widest possible audience. If the news is likely to be of interest to many people—about a major company expansion, or an important new product—then the release goes to local newspapers, to the wire services, and to national business newspapers and general circulation magazines. And it will certainly go to the technical press.

Now, no release can meet the needs of all these publications. Each publication must take the release and shape it to fit its own editorial scope. As it happens, it's the technical press that has to do the most rewriting when the release is general, for such releases will no doubt contain little that a technical audience doesn't already know. Also as it happens, this isn't bad, for the technical press is in a position to do the best rewrite job anyway.

Even where a company designs a release for the technical press itself, it will send it to several different publications. In the chemical industry, this means perhaps half a dozen magazines serving the entire chemical industry and perhaps a like number serving a vertical segment of the industry. Again, no release can serve all these publications equally well, so each editor should rewrite it to serve his own readers better.

READERS WANT SIGNIFICANCE

By rewriting, an editor performs his greatest service. At one time, publications contented themselves merely with recording facts. Today, that no longer is enough. Industry is enormously complex; events occur at a rapid pace; today's new product is tomorrow's has-been. Moreover, the average man in industry is flooded with technical material to read. He is most interested in covering this material quickly. Above all, he wants the significance—what it means to him.

Filling these needs is where the editor comes in. He takes the basic information in technical news releases that tells who does what and where, and he adds to it—searches his files, looks back at other stories he may have published in the same field, calls the company on the phone, or interviews people there in person to find out more on the background. Then he puts it all down on paper, so that the reader can make the most use of the information gathered.

REWRITING

How you rewrite specific releases depends on the subject, the magazine, and your purpose. There are no hard and fast rules to follow. But let's look at two specific releases to see some of the things you can do to fill them out. The first release is on a production process, the second on a new company and plant.

Largest Synthetic Mica Melt Completed

The largest commercial melt of Synthamica synthetic mica yet attempted has been successfully completed by Synthetic Mica Corporation, a subsidiary of Mycalex Corporation of America, at its plant in Caldwell Township, N. J., according to an announcement by Jerome Taishoff, president. Over 80,000 pounds of raw material were used.

After approximately three weeks of cooling, the huge mass of crystalline mica will be broken up and then ground or split into usable form. "It is too early to tell what progress has been made in the growth of large area crystals in this melt," Mr. Taishoff said, "but each melt we make has produced bigger and better crystals, and we expect that the progress will continue." Crystals as large as 4 by 4 inches have been produced consistently in the past, and intensive research is making steady progress in the production of large area crystals in greater volume.

Investigations of the processing of Synthamica synthetic mica for industrial use are proceeding along two general lines: the growth and mechanized splitting of large area crystals; and the development of a reconstituted sheet of mica platelets without the use of chemical binding agents, the latter course being pursued under a two-year contract with General Services Administration. Developments are being watched with interest by various government agencies and manufacturers of high-temperature electrical and electronic equipment, since synthetic mica is superior to the natural material for applications involving temperatures in excess of 1000°F. Experiments in the control of furnace operations and cooling to produce large crystals are currently being conducted as its own responsibility by Synthetic Mica Corporation.

Synthetic Mica Corporation was established as a wholly owned subsidiary by

Mycalex Corporation of America in 1955 to further the commercial development of synthetic mica. The synthesis of this strategically important material had been brought to pilot-plant production in Germany during the war, when the Axis countries were cut off from sources of natural mica in India and elsewhere, and found their electrical and electronics development severely curtailed as a result of the lack of material. Foreseeing similar possibilities in the future, the United States Government, through the U.S. Bureau of Mines and various industrial participants, initiated a program at Norris, Tenn., for the commercial production of synthetic mica. In 1953 Mycalex Corporation of America entered into a cooperative agreement with the Bureau of Mines to produce the material, and at the conclusion of the agreement built plant facilities and became the only commercial producer of Synthamica synthetic mica.

That's a good release. It tells right at the beginning that one company has made the largest synthetic mica melt, how much, what will happen next. It goes on to tell why the company is doing this—to make larger and larger single crystals. Then we have a paragraph on processing synthetic mica, one on who's interested, and a final one on this company's background.

It's short, and to the point, but it can stand amplifying. First, you must decide how you will slant the story and still have a news story. With this release as a starting point, you could amplify it to make a story on research in synthetic mica. Or you could write on mica as a material—

where it is used. Another possibility would be a production story—how you do make it. Each of these you could tie to the specific news event of the largest single melt.

Let's take the production approach as the example. The key questions for amplifying this release, then, would be, "How do you make synthetic mica, and what are the various problems you face?" With these as starting points, you can then round out the story: "Why are we interested in making synthetic mica?" "Who developed the various processes?" And finally, "How about more information on where these products are used?"

Here, then, is the story that developed after the company answered these questions and after the editor picked up other information from various source files:

Synthetic Mica Moves Ahead*

Picking up where Germany left off, government and industry develop better process and product.

Last week, Synthetic Mica Corp. of Caldwell Township, N. J., began to break up its largest melt yet of Synthamica synthetic mica—80,000 pounds worth. Successful completion of a melt this big focuses attention on this country's slow but steady progress toward independence from foreign sources of mica.

People have been thinking about synthesizing mica for 100 years or more. But American interest dates back to 1945, when technical intelligence teams invaded Germany to study its scientific progress during World War II. These teams found that the Germans had brought a process for making synthetic mica all the way to pilot plant scale and had already made one mica crystal six to eight inches square.

The teams recommended that this country take up the problem. In 1947, the Bureau of Mines did so, spurred on by a desire to have a substitute for natural mica should our supplies be cut off in any future wars.

---

* Reprinted from *Chem. Eng. News,* **35,** No. 31, 110, 134 (1957).

BuMines has worked on the problem since that time. It developed a process using internal resistance of the melt for heat and using the raw materials to contain the melt. It made sheet-mica crystals as large as 2 to 3 inches on a side from melts of up to 10 tons. It helped to develop a process for reconstituting smaller particles into sheets—necessary since yield of large crystals per melt is so small.

In 1955, Mycalex Corp. of America took over commercial development of the product, forming Synthetic Mica Corp. to handle the assignment. Since then, Synthetic Mica has continued to improve on the foundation laid by BuMines. It now makes crystals as large as 4 by 4 by ⅛ inch routinely, although still in low yield. And it melts larger and larger batches and is now up to 80,000 pounds—its largest yet.

"It is too early yet to tell what progress we have made in growing larger crystals in this melt," says Synthetic Mica president Jerome Taishoff. "But each melt we make has produced bigger and better crystals than the one before."

How To Make It. To make synthetic mica, the company starts with a mixture of silica sand, calcined alumina, calcined magnesia, potassium silico-fluoride, and an auxiliary source of potassium such as feldspar.

These ingredients are mixed in proportions close to the one desired in the product, except they are richer in fluorides and potassium to allow for losses in processing. They are heated in an electric arc furnace up to 1300° to 1400° C., and held for hours, melting 40% to 50% of the crudes.

When melting has progressed to this point, the mass is cooled slowly—a degree or less per hour—back to room temperature. Cooling rate is one of the most important factors leading to large crystals, the company says.

The resulting synthetic mica is of the fluor-phlogopite type ($K_2Mg_6Al_2Si_6O_{20}F_4$). Synthetic Mica prepares this form rather than muscovite—the natural form used widely in electric and electronic equipment—since its synthesis is easier than that of muscovite. Usually, there are only a few large crystals per batch; those that can be separated from the mass easily are removed and used separately—for vacuum tube spacers and the like. The remainder of the melt—mostly small crystals of synthetic mica—is crushed, ground, mixed with glass and other ingredients, and molded into final shapes.

As the company has gained experience, it has been able to

make dielectric products—marketed under the name Supramica ceramoplastic—which can operate at higher and higher temperatures. It now has one which, it says, can operate continuously at temperatures up to 500° C.

Supramica products are already in use in missile guidance systems and radomes, as well as in many electric and electronic components and in jigs and dies in dielectric and induction heating applications. Synthetic Mica points out that mechanical properties, too, offer some applications—it can be mixed with metal powders or wires to form cermets and can be used as a filler for plastics, waxes, and paints. It can be combined with abrasive grain to make grinding wheels. Uses other than for its electrical properties are mostly still in the talking stage, though.

What next? Synthetic Mica is still improving its process and its product, working on two fronts:

• Growing bigger crystals and more of them, and then separating them from the melt by an economical method. Large crystals are now separated from the cooled melt by hand picking, a time-consuming and costly method.

• Developing a reconstituted mica sheet that does not use chemical binding agents—such as the glass now added to Supramica. Pure mica sheet has better dielectric properties than the mixture.

First, notice the new lead. The original lead sentence said "The largest commercial melt of . . . synthetic mica yet attempted has been successfully completed by Synthetic Mica Corp. . . ." The new lead emphasizes an important point in "editing" news releases—give the reader a feeling something has just happened: "Last week, Synthetic Mica Corp. . . . began to break up its largest melt yet of . . . synthetic mica—80,000 pounds worth." From a rather pedestrian start, we now give the reader instead a revised lead that points out the timeliness of the event.

Now the second sentence: "Successful completion of a melt this big focuses attention on this country's slow but steady progress toward independence from foreign sources of synthetic mica." And that sentence sets the scene for the balance of the story—why this development is important.

From there, the story as published goes on to answer the questions mentioned before, sometimes using material in the release, sometimes material gathered in a phone call, sometimes material from magazine source files: When the U. S. became interested in synthetic mica, what part the Germans played, how the Bureau of Mines took over. Then a bit on Mycalex Corp. of America forming Synthetic Mica Corp. (this from the release), how this company has gotten crystal sizes up to 4 by 4 by ⅛ inch routinely.

Since this is mainly a production story, the story has some details on the production process itself, a paragraph or two telling where synthetic mica fits in industry, and a bit on what the company expects to do in the future. All of these points the editor has added.

You can see that this story now goes much farther than the original release, and gives readers much more information than they would have gotten from the release alone. Not that the release was poor; it wasn't. But the rewrite is a good example of how an editor can build a technical news release into a better story for his audience.

Let's look at another release, this one telling about two companies forming a joint company to make high-purity silicon:

New York, July 15—Two century-old firms, W. R. Grace & Co. of New York and Pechiney, giant French chemical and metallurgical concern, announced today the formation of a new company in the United States to produce high-purity elemental silicon and other semi-conductors.

Silicon, which is produced from one of the earth's most abundant raw materials, sand, is finding large scale use as a semiconductor in transistors and rectifiers for highly sensitive electronic devices.

Grace and Pechiney anticipate that the plant, the location for which has not yet been announced, will be operating within less than twelve months, and that annual capacity will reach 20,000 pounds. Demand for silicon to be used as a semi-conductor has been increasing rapidly. Pechiney is already producing highly purified silicon in France, and under an arrangement between the two companies, the new corporation will be selling this imported material through International Selling Corporation in the United States within thirty days.

J. Peter Grace, President of W. R. Grace & Co., said, "This is not the first venture that Grace has undertaken with Pechiney. In 1952 we decided to enter the chemical industry as manufacturers of ammonia and urea in Memphis, Tennessee and in that plant we were one of the first American firms to produce urea using the Pechiney process. This plant is currently exceeding rated capacity (55,000 tons per year)."

Mr. Grace continued, "Since 1954, we have been importing high-purity aluminum from Pechiney for our Davison Division's reforming catalysts used in the production of super octane gasolines."

He added, "The arrangement under which the new company will import Pechiney's French silicon pending completion of our new plant gives us the advantage of having this product to sell almost immediately. The advantages of the arrangement include know-how agreements covering all types of semi-conductors that Pechiney might develop."

The program which led to the formation of this new company was initiated by Grace Research and Development Company Division, one of the seven chemical divisions of W. R. Grace & Co. The new company will be under the director of Grace's Davison Chemical Company Division, already concerned with the production of other rare metals such as thorium and uranium.

The use of semi-conductors in transistors started with the manufacture of germanium transistors in 1950. Production of high-purity silicon, in which impurities are measured in parts per billion, was initiated in 1954 in the United States.

Today's pocket radio and hearing devices are made possible by the use of these transistors, invented by Bell Telephone Laboratories in 1948. Other uses for silicon devices are in electronic instruments for defense such as electronically-controlled weapons and in protective devices in supersonic flight, guided missiles, undersea warfare, and the solar energy battery.

Transistors are also used in high fidelity equipment, tape recorders, dictation instruments, television sets, and in automatic devices such as computers and controls.

Silicon rectifiers may also be used to

convert electrical systems in automobiles from alternating current to direct current, effecting savings in equipment and imparting longer life to car batteries.

Germanium and silicon are the only two semi-conductor materials of importance today. Domestic sources of germanium cannot supply the demand. Silicon is twice as costly as germanium. However, the abundance of raw material supplies and its ability to withstand temperatures 100°C. higher than germanium make it a leading contender in the mushrooming electronics market.

Again, a good release. It tells who is involved, what they will do. It tells that the plant will make 20,000 pounds a year and be running in less than 12 months, that Pechiney already makes the material in France, and that this is not the first joint venture. It ends with five general paragraphs on where silicon goes today.

But it can stand rewriting. Look at the second paragraph, for example: "Silicon, which is produced from one of the earth's most abundant raw materials, sand, is finding large scale use as a semi-conductor in transistors and rectifiers for highly sensitive electronic devices." True enough, but hardly news to readers of the chemical industry press. The same can be said about the final paragraphs telling about silicon's place in electronics. Here's where the technical editor comes in, to leave those familiar parts out and instead put in additional information that will interest his readers:

• Who are the other producers of this material?

• How has the price changed for this material with the rapid expansion that's gone on in recent months?

• U. S. and foreign companies have formed a lot of joint

companies lately. This isn't the first joint one between these two, but will there be more? If so, in what fields?

• Where will the plant be? The release says location hasn't been decided, but that doesn't keep a good editor from asking. With these answers, we end up with a story that looks like this:

Semiconductors in Hot Race*

ANOTHER FIRM ENTERS the electronics raw materials business. W. R. Grace and France's chemical giant, Pechiney, last week revealed they would form an as yet unnamed company to produce high-purity silicon (C&EN, July 15, page 7). The company, to be run by Grace's Davison Chemical division, will go on stream in less than a year. Its annual capacity of 20,000 pounds will be used largely as a semiconductor in transistors and rectifiers.

Plant location will probably be in Puerto Rico, unless atmospheric tests now under way prove unfavorable. In that event, location would be at one of several mainland sites Grace is considering. Puerto Rico would provide advantages; officials point out that relative remoteness is not of economic importance with output in pounds rather than tons. Grace already has Puerto Rican facilities, and is currently considering a paper mill there.

This is Grace's initial joint venture with a foreign chemical company in the United States. However, it has a few similar agreements outside of this country, particularly in South America.

Why Pechiney? Grace officials answer that in several ways. First, they know the company; the two began dealings five years ago when Grace commenced negotiations to produce urea via the Pechiney Process at Memphis, Tenn. Pechiney supplies high-purity aluminum to Davison Chemical for reforming catalysts, and has since 1954.

Since 1951, prior to joining the Grace organization, Davison had been working jointly in France with Pechiney on various petroleum cracking catalysts. Davison, of course, is the division in the picture because of its background in high-purity metals.

---

* Reprinted from *Chem. Eng. News*, **35**, No. 29, 22, 23 (1957). Copyright 1957 by the American Chemical Society and reprinted by permission of the copyright owner.

But most important, Grace officials think a lot of Pechiney's silicon product. Pechiney is now expanding and plans to market silicon here through the new company within 30 days (it is now sold through an affiliate, International Selling Corp.). High-purity silicon is subject to a 12.5% tariff—one strong reason Pechiney wants to produce it in the U. S.

Officials are not overly concerned about increasing competition in this facet of the electronics supply industry. In addition to Du Pont's Brevard, N. C. facilities, Sylvania, Eagle-Picher, and Texas Instruments are in production; and Mallinckrodt has high-purity silicon in its development program but has no production plans at present.

Charles E. Waring, vice president of Grace Research & Development, expects market demand to be for quality rather than price. Depending on quality level (based on resistivity) high-purity silicon prices generally begin around $320 per pound, ranging up to $900 per pound for research grade material.

Right now, Grace has no specific plans to expand into other parts of the electronics industry. M. G. Geiger, W. R. Grace executive vice president in charge of the chemical group, notes that the agreement covers all materials in the field Pechiney may develop, including other semiconductors. While there is now no basis for the view, this operation may lead to other joint ventures. It could provide a process know-how and patent exchange agreement in the future. Pechiney now manufactures vinylidene chloride monomer in France, which might prove an impetus for Grace's Cryovac division.

As Waring points out, "We expect to be doing business with each other over the years. With common interest both here and in Europe, the future could bring much cooperation."

Notice the new lead: "Another firm enters the electronics raw materials business." It sets the scene immediately with a sentence that tells the reader "here's a trend." Then, "W. R. Grace and France's chemical giant, Pechiney, last week revealed they would form an as yet unnamed company to produce high-purity silicon. The company, to be run by Grace's Davison Chemical Division, will go on stream in less than a year. Its annual capacity of 20,000 pounds will be used largely as a semiconductor material in transistors and rectifiers." And there you have most of the meat right at the beginning: Who, the time element "last week," who will run the plant, its capacity, when it will start.

Then the article begins to bring in supplemental information. "Plant location will probably be in Puerto Rico unless atmospheric tests now underway prove unfavorable. In that event, location would be at one of several mainland sites Grace is considering. Puerto Rico would provide advantages. Officials point out that relative remoteness is not of economic importance with output in pounds rather than tons. Grace already has Puerto Rican facilities, and is currently considering a paper mill there." And further along in the story we come across this: "High-purity silicon is subject to a 12.5% tariff—one strong reason Pechiney wants to produce it in the United States."

Then we have a paragraph of additional background on who's in the silicon race: "Officials are not overly concerned about increasing competition in this facet of the electronics supply industry. In addition to Du Pont's Brevard, N. C., facilities, Sylvania, Eagle-Picher, and Texas Instruments are in production, and Mallinckrodt has high purity silicon in its development program but has no production plants at present."

You can see this a much better news story for having been rewritten than it would have been had the straight release been used. This one, then, and the earlier story on mica show what editors should do to bring in added information, to weed out the unnecessary, and to interpret the information for their readers.

WRITE SIMPLY; WRITE CONCRETELY

One part of technical editing, then, is rewriting to bring in supplemental information. The second part is *how* you write. As mentioned, people are busy today. They want their information quickly. They want it in a form that's easy to understand. But what has happened to technical writing today? Too much of it is not easily read. We take complex subjects and then proceed to discuss them in the most tortuous way. Too often when a man takes typewriter in hand he trots out what he feels is his top vocabulary and writes to *im*press, not *ex*press.

You'll find that sentences become long and involved. For example: "Dr. A. T. B. and Mr. R. M. of A Co., conceivers, designers, and producers of the world's first mass produced portable nuclear reactor, announced today that as a further extension of A's services and in an effort to alleviate technical manpower shortage in the nuclear technology field, they are offering free of charge to a limited number of university and college educators the opportunity to attend A's progressive reactor training program in reactor theory and application." Try to understand all that at the first reading—or to read it very rapidly.

Take another pitfall—noun writing. Get oriented to verbs, because verbs mean action, and action makes for interesting reading. Instead of "for the manufacture of," use the simple "to make." Instead of "on the source and mode of preparation," try "where it comes from and how it is made." That's more words, but it's also easier to read. For "the arrangement of atoms within the molecular structure," how about "how atoms are arranged in molecules."

The flag to stop you here and warn you that there are probably better ways to get the same thought across: Are too many words ending in "-ent," "-ment," "-tion," "-able," and all the other suffixes we tack on to change verbs to nouns? If so, rewrite with action words.

Look out for unnecessary words or jargon; we've got a lot of both in technical writing. How easy it is to sprinkle writing with words that don't carry their own weight: fundamentally, essentially, basically. Or use nice fat redundant phrases—"hexical structure" instead of the simple "helix." Or, the "in" phrase—round in shape, green in color—as if it might be round in size and green in taste. Another: "carry out detailed studies." Why "detailed?" Most likely to impress the reader, but if the study wasn't detailed, then maybe it isn't worth mentioning.

One final point: Steer clear of the abstract word. Forget "characteristics," "facilities," "factor," and the like. Use concrete words instead. For instance, "AEC requests notification of accidents (with radioactive materials) so it can minimize the consequences." Now anyone reading that will wonder right away what "minimize the consequences" means. Will AEC get out fire hoses and sluice the stuff down the drain? Maybe it will merely prepare a report and then classify it so Congress won't find out. As it turns out, the phrase meant AEC would send an inspector to the scene to make sure people involved would use the right clean-up methods. That's hardly apparent in "minimize the consequences," however.

Anyone who deals with technical news releases will have to guard against these failings like the plague, because this is the way too many releases are written. Your writing becomes dull and heavy. When it does, people stop reading it, and you've failed in your main goal—selling information.

Let's look now, at some examples of rewriting that show how you can overcome these problems of long sentences, noun writing, and words that mean nothing:

> The fundamental basis of the coal chemical industry is essentially the relation between costs of B.t.u. from petroleum and the price obtainable for refined distillates from coal carbonization. The steel industry now buys much liquid and gaseous fuels which, in principle, might be largely spared in part by burning its own tar.

A bit complex, what? Notice the jargon words "fundamentally" and "essentially." See how verbs have changed to nouns—basis, relation, obtainable, carbonization. Also, note the weaseling: "largely spared in part." Here's the rewrite:

> Just how much coal chemicals the steel industry produces depends on the price it must pay for B.t.u.'s from petroleum and the price it can get for refined distillates from coal carbonization. The steel industry now buys much liquid and gaseous fuel it could side step—in principle at least—by burning its own tar.

Five words longer, but easier to read.

Here's another example, carrying the verb-for-noun business farther and also showing how to change abstractions to something concrete:

> Children are inquisitive, but this inquisitiveness seems to stagnate during adolescence. To develop the minds of our best students we must show them the satisfaction they can obtain by delving beyond the limits of the usual high school curriculum.

The revision:

> Children ask questions, but by adolescence they seem to stagnate. To keep our best students at it, we must show them how much they can enjoy going beyond what the usual high school hands them.

Notice "Children ask questions" brings a mental picture to mind; "Children are inquisitive" falls rather flat.

Finally:

> A factor partially responsible for ability of the corn plant to resist European corn borer has been isolated at University of Wisconsin. By determining the amount of the factor present in a given line of corn, a plant breeder will thus be able to breed plants which will

> be resistant to insects. The material has no insecticidal activity, but Smissman and associates are studying other, potentially more active compounds in the series.

The rewrite:

> A chemical that helps corn resist European corn borer has been isolated at University of Wisconsin. The material won't kill insects itself, but Smissman and associates are studying other more active ones in the series that may. Also, by knowing how much of the material is in a given line of corn, plant breeders will have a better idea of how they are getting along in breeding resistance into corn.

Notice how we get rid of the abstract "factor," substitute "chemical." Then there are verbs in there disguised as nouns—ability, resistant, activity.

In short: be concise and concrete. Prefer the simple word to the complex and the simple sentence to the long. You'll find that your writing will be clear. Your readers will find it interesting.

Most readers will, that is, for there's still a die-hard core of people who feel simple writing isn't dignified or professional. Nonsense. Pick any successful magazine or book written today, and you will find it will be written simply and concisely. As one publishing company puts it, "The ideal article (is) written so that the words are for children and the meaning is for men." Exactly. Or, as one speaker on this subject once said, "Who would have remembered Ben Franklin had he said 'Early to retire and early to arise makes a man salubrious, opulent, and sagacious..' "

Part III

EDITING BOOKS AND MANUALS

Chapter Twelve

Publisher Editing of Technical Books

G. G. HAWLEY
Executive Editor, Reinhold Book Division
New York, New York

A highly effective description of a technical book editor was offered some years ago by a person of considerable eminence in the technical publishing world. "A good editor," he said, "is like a river which is a mile wide and an inch deep." Those who are not quite as pessimistic as the originator of this simile might amend it to the extent of increasing the depth of the river to two inches! In point of fact, however, there is probably no editor worth his salt who does not at least once a day privately lament the superficiality of his own knowledge of the broad array of subject matter with which he is required to deal. And he is right in so doing. For the technical editor, when he is functioning responsibly, must be more than a coarse screen to remove large bits of undesired matter from the stream of material that flows through his office: he should take an active part in shaping it to the needs of its readers and in winnowing the wheat from the chaff, which sometimes pops up in the most unexpected places. It is obvious that if an editor is able to do this well, his knowledge must be broad, and also as deep as possible. Of the two dimensions, the breadth is far more important, for in cases where detailed specific knowledge is necessary the editor can rely on his library and other sources of information.

EDITORIAL LEVELS

Having thus characterized the professional editor as a shallow individual, we shall outline the various levels at which the editorial function is carried out by technical-book publishers. There are three of these levels, and it must be borne in mind that they will vary to some extent with the nature of the books involved and with the policy of individual managements. For example, publishers of college textbooks have editorial problems peculiar to their industry; so do publishers of high level treatises, reference books, and handbooks. With due regard for these differences, the three types of editorial function to be described are generally applicable to all kinds of technical books. The nature of the activity of each level will be discussed and the necessary personal characteristics and qualifications suggested.

Top Level

At the highest level of operation, the editor works in liaison with his management. His duties here are to develop a specific policy as to the best kind of books for his company to publish. Frequently, and quite naturally, top management will have its own ideas, behind which strictly economic factors usually lurk. None the less, wise management will not only urge, but require its executive editors to recommend general publishing policies. Hence, it behooves any top-level editor to familiarize himself with at least some of the financial facts of life and to acquire an appreciation of management's problems. Editors can rarely be ivory tower dwellers—at least in the technical field. The question of what types of books are best for a given publisher is too complex to discuss here; but it is certain that one of the most valuable contributions an editor can make to his firm is to study its special situation and problems and the history of its activities.

Once the over-all policies are agreed upon, the alert editor will try to conceive and develop books which best embody them. It is sometimes (but not invariably) true that the most successful technical books originate in ideas propounded in the publisher's own office. Here again, management will look to its editors for constructive project suggestions. It then becomes the editor's duty to find suitable authors for the books agreed upon. This function is frequently designated by the rather dubious term "procurement."

Believe it or not, procurement is a major job in itself. In some companies, it is performed by one or more field editors reporting to the editor-in-chief. They spend a great deal of time on the road, attending meetings, and calling upon academic and industrial prospects. Textbook houses employ a large staff of so-called "college travelers," partly to acquire new manuscripts from professors and partly to urge them to adopt recent texts in their courses. In both cases, the responsibility for implementing the publishing policy usually devolves upon the top editor.

Still another function of the highest editorial level is to decide upon the acceptance or rejection of book ideas which originate *outside* the company. This is usually done in conjunction with the sales and promotion departments, and often with the production department as well. Such decisions are hammered out in editorial conferences or panel meetings, the factors being marketability, price, manufacturing cost, and general suitability to the publisher's list. The editor must be thoroughly conversant with sales-promotion methods, production problems, and the competitive situation in order to function effectively in this important phase of his work.

Some of the qualifications and personality traits necessary for the top-level editor may be readily inferred. First, of course, is a fairly thorough knowledge of the areas in which his company's interests are uppermost. He should have a reasonably good background of technical training and experience, and have the facility of acquiring new knowledge rapidly as he goes along. Even at best, his knowledge of a specialized field will rarely be more than "an inch deep" compared with that of a recognized authority in this field.

Therefore, other characteristics than knowledge are necessary—for example, ability to meet people of high intellectual standing and to discuss book problems with them intelligently and sympathetically. In dealing with experts, such qualities as tact and astuteness are even more important than spot knowledge. A good editor never makes the mistake of arguing with an authority about his specialty, or of making him look second-best at a dinner-table discussion.

On the other hand, the editor should avoid cultivation of a "smooth" approach, which often arouses distrust and implies lack of sincerity. The hard-hitting "sales pitch" personality is almost equally obnoxious, particularly to academic men. The essential qualities are dignity, poise, and above all an understanding of the prospective author's problems, whether they seem trivial or abstruse. Display a genuine interest in his ideas, his hobbies, and even his personal life. Praise his wife's cooking when you are invited to dinner. Learn how to listen intelligently. All these are invaluable assets, and it is surprising how effective they can be. On the other hand, it is a mistake to get too intimate with prospective authors, as this may backfire in an embarrassing way if trouble develops with the manuscript.

It should be pointed out that sometimes all three types or levels of editorial activity must be carried out by the same individual. The smaller the company, the more detailed the responsibility which falls upon its editor. As soon as a book-publishing operation emerges from the infancy class, the three levels are appropriately divided; for example, there will be an editor-in-chief, who is supposed to be capable of operating at *all* levels; reporting to him will be one or more "procurement" men or college travelers; and also working under his direction there will be several assistant or junior editors (usually female). Although the top-level editor is usually not directly responsible for physical production of books, dealing with printers, etc., he is responsible for keeping a continuous supply of edited manuscript flowing into the production department. Sometimes he may have a first assistant to whom are assigned the duties of the second level, which will be discussed shortly.

To summarize, then, an editor operating at the highest level should be reasonably intellectual, interested in both things and people, sincere in a desire to benefit science in general as well as his own and his company's interests, and creative in his attitude toward his work.

Second Level

The fact that an author has been found and a contract signed for a given project by no means guarantees its completion. One of the most important of the second-level editorial activities is that of checking up on and needling would-be authors to complete their work. Often they need encouragement, guidance, and even active collaboration. Sometimes they require real prodding. The editor must decide in a given case whether to wheedle and coax, aid and encourage, or apply the spurs. Seldom indeed can he afford to "get tough," no matter how justly he is provoked; for he cannot compel an author to fulfill his contract, and unwise criticism may cost him the book. It is a remarkable fact that there are some scientists of unimpeachable character and position who not only fail to fulfill contractual obligations, but do not even bother to advise their publishers of their intentions. This situation creates a serious problem for the editor who is responsible for maintaining a publishing program, and also for the management.

Besides prodding delinquent authors and contributors, the second-level editor must evaluate the completed manuscripts which he does receive. The main factors here are: (1) accuracy of subject matter, (2) organization and presentation, (3) coverage of subject, and (4) English and readability.

For review of subject matter and coverage, the manuscript is usually submitted to a referee of the editor's selection. If one or two adverse reviews are received, the material is returned to the author for correction. If the chief trouble is in the organization and presentation (here the editor's judgment may be better than a referee's), a decision must be reached as to whether the weaknesses can be taken care of by routine editing, or are serious enough to require that the manuscript be rewritten or even rejected. Sometimes the author can find a ghost writer among his students or friends; sometimes the editor finds one for him. If the editor has the necessary time and knowledge, he may undertake the job himself.

In some cases, the editor will be forced to reject an inaccurate or poorly written manuscript—indeed, this is a prime responsibility, however distasteful it may be. Any editor who takes his work seriously regards himself as the protector of his company's interests and also of those of his readers. He can ill afford to allow a book to appear which will be severely criticized by reviewers or bring threats of legal action. He must be incessantly alert for statements not founded on fact, or for strongly biased conclusions. He must pick these up and have them out with the author; if this duty is neglected, serious trouble may result.

It is evident that we have here a situation in which the editor may find himself in a dual position. Whether the first and second-level duties are performed by two men or by one man is a minor matter. From a functional point of view, successful operation demands that the editor must first act essentially as a *salesman* (when he is trying to induce an author to write a book or to obtain for his company a manuscript already completed), and secondly as a *critic.* It is always a temptation for an editor or procurement man to overcommit himself to an author, perhaps with the thought in mind that "after all, once we have him signed up we can cut out the bad spots later." Because of this duality of function the experienced editor will be somewhat less acquisitive to begin with, for he may have had the horribly embarrassing task of rejecting or strongly criticizing the work of a man whom he had been assiduously cultivating a few months previously.

The attributes required for the second level of book editing are suggested by its duties. Obviously, they include qualities very different from those previously described for the highest level—namely, a rather intense skepticism, a more than rudimentary critical capacity, and a feeling for reader appeal. In short, a good editor at this level should be able to distinguish a good book from a bad one. Often, an excellent field editor or procurement man makes a poor evaluator. An editor who must operate at both levels simultaneously (and many do) must be something like the country-hotel clerk who puts on a different cap for each of his several jobs.

Third Level

After a manuscript has been written, submitted, criticized, revised, and finally accepted for publication, it descends into the clutches of the third rank of editorship—which is in its way quite as important as the others. Work at this stage (which might be called the "busy bee" level) is usually carried out by assistant or junior editors—often young ladies of good education and ability who like this sort of activity. A senior editor is virtually helpless without two or three such assistants on his staff; their acumen and devotion to their craft are often most admirable.

They perform a variety of services which may be summarized as "preparing the manuscript for the printer." These include the following:

1. Editing and scaling illustrations. Many photographs are too poor in quality to reproduce well. Line drawings vary widely in size, lettering quality, and general reproducibility. The editor must review this situation, returning poor copy to the author and having photographs and line drawings relettered or retouched by the art department where necessary. He or she must also see that captions are correctly applied and that proper reference is made to the illustrations in the text. Location of pictures must be indicated on the galley proofs to guide the printer in page makeup.

2. The assistant must also read the manuscript in detail for errors in English, punctuation, spelling, and, above all, general consistency. Here a "nose" for factual detail and foggy, meaningless sentences is essential. Often editors with little or no formal education in science are able, by sheer common sense and ability to observe and reason, to find serious errors that sometimes escape not only the author but an entire reviewing staff as well. This does not mean

that an editor in working over a manuscript has to investigate every technical detail; but it does mean that an occasional spot check sometimes pays off. One is usually led to make such spot checks by a queasy feeling that all is not well with the manuscript. This arises from unpredictable details such as the misspelling of a common name like Priestley, putting the island of Aruba in the Indian Ocean, or giving the boiling point of water as 212° Centigrade. Such things happen, and happen frequently.

3. The assistant editor must, of course, refer such matters to the author, and check over with him her changes and other queries. Such close liaison with the author usually pays dividends in good will and improved books. Where disagreement persists, the matter is then passed up to the chief editor, and he decides it one way or another with the author.

4. One of the most tedious (and at times frustrating) phases of the assistant editor's job is to see that consistency is maintained throughout a manuscript, as regards such minutiae as abbreviations, spelling, reference style, table arrangement, and general technical notation. This is an especially difficult task when working with multi-authored symposia, for each author seems to use his own scheme, and usually the editor in charge of assembling the material is too much occupied with the technical unity of the book to pay much attention to these details. In cases where chapters are contributed by foreign authors, the variations in spelling and chemical nomenclature must be reconciled to standard American practice. Sometimes there is lack of consistency in use of temperature scales. Some chapters come through with the arithmetical notation used in Europe—that is, commas for decimal points and periods for commas in setting off digit groups. It seems as if there are as many variations in the style of presenting bibliographic references

as there are authors. One sometimes wonders if there is any possible variation that has not been used—and usually it will turn up in the next 24 hours!

5. Besides all this, the editorial assistant must "style" the manuscript for the printer; that is, indicate what sizes and weights of type should be used for chapter titles, section titles, main heads, subheads, sub-subheads, and so on. Often this is complicated by failure of the author to maintain a logical order of topical breakdown, and it is often necessary to refer such matters to him. When he is not easily available, the chief editor may have to take a hand. It is advisable to require authors to submit topical outlines with the manuscript; but in cases of multi-authored books this is seldom possible. In two chapters of approximately equal length, sometimes one will have no subheads at all and the other as many as two or three on every page. Here again, a consistent and unified system must be decided on and followed.

Perhaps enough has been said to give a rough idea of this level of editorial work. It requires a good memory, a certain sense of skepticism, and a readiness to use reference books liberally. The more dictionaries and data books one has at hand, the better. One must also have a natural liking for words and their meanings, and an innate feeling for consistency. Good editors, in this sense of the word, are born rather than made—a fact which is generally not understood by the many aspirants to this profession.

BOOK EDITORSHIP AS A PROFESSION

In a very real sense, the high purpose of any occupation is, or should be, to derive therefrom a sense of value to society and satisfaction of innate personal desires. Some men design bridges and skyscrapers, some become research scientists, some business executives, and some physicians, depending on their individual predilection and ability. One of the deepest instincts of man is creativeness, and any occupation which excludes this factor can be no more than a temporary stepping-stone for the normally ambitious person. But it is important to note that what one man will regard as a means to an end, another of different temperament and talents will strive for as the end itself. An editor gets a sense of creativeness—of fulfillment—when he looks at a finished book with the realization that its existence in that form is largely due to him, because he either conceived it or contributed so effectively to its execution that it may be said to reflect his own ability and workmanship. Admittedly, such rewards are only occasional, but their possibility is an ever-present source of comfort.

Of the professional qualities of technical editorship, Dr. William A. Hamor, Senior Director of Research of Mellon Institute, has this to say:

"The technical editor is a powerful force for infusing science in the professions and in technology. It sometimes appears to me that there are so many kinds of editors that any one can call himself an editor, and that this situation is a social ailment. . . . Every editor worthy of that title must apprehend ethical meaning—the discriminative capacity between right and wrong. Editorship of proper quality really constitutes a profession. . . .

"The most important asset of an editor is to be able to make decisions having soundness. And this characteristic implies that he is also able to assume and exercise responsibility. In his relations with others, with the specialists in his own organization and with potential authors, the technical editor must be sensitive to human traits and reactions, and must have a high degree of integrity, sincerity, and fairness."

Editorship right down the line requires knowledge, intellectual capacity, judgment, tact, and ability to accomplish necessary results without offending people—all of which distinguish a profession from a job. A profession requires both maturity and responsibility. The editor's personal credo can be condensed into two words: *professional responsibility*—and this applies to all the levels described. An editor is responsible to his readers, to his management, and, more important than either, to the truth. If he can achieve all this, he is indeed a full man.

Chapter Thirteen

Editing Technical-Equipment Manuals

SAMUEL A. MILES

Vice-President, Miles-Samuelson, Inc.
New York, New York

The editing of technical-equipment manuals must be approached from a fresh point of view which reflects the rapid changes in our technological progress. Such an approach is necessary because the demands occasioned by our commercial and military advances have placed a great burden on all people concerned with the delivery of accurate, up-to-date, easy-to-use information in the complex fields involved.

To be certain that we are all thinking about the same things, I would first like to explain what we mean by technical-equipment manuals and how editing them differs from more-conventional "technical editing." This distinction is important because the high degree of specialization involved in editing technical-equipment manuals makes it likely that many of my remarks will not be altogether applicable to editorial problems outside this special field, or even outside the field of military technical-equipment manuals.

TECHNICAL-EQUIPMENT MANUALS

Many thousands of technical-equipment manuals have been prepared, and new ones are being worked on every day. I can best explain the term "technical-equipment manual" by describing one type of manual which encompasses, because of its complexity, practically all the different elements which may be encountered. I am referring to the manuals which are part of our important guided-missiles program.

Function of Technical-Equipment Manuals

Accurate delivery of the payload is the sole purpose of our military missiles program. Because the weapons systems designed to do the job are so complex, we often lose sight of this simple fact. Furthermore, we are beginning to realize that a weapons system will not deliver the payload unless trained people are in command of the missile. At the launching site, therefore, we must have a package containing both the equipment and the trained personnel. The two components of the package—missiles and men—have many, many supporting items. The supporting items are themselves complex, and require for their successful operation the services of men who are well trained in their special spheres of activity.

Training is as old as the human race. A primitive "technical writer," if he were so inclined, could have produced on a cave wall all the elements of a "technical-equipment manual" (see Figure 13.1). This primitive manual has evolved into a complex series of technical-equipment manuals, training manuals, job cards, check lists, training films, film strips, charts, etc. In this chapter, however, I shall confine myself to only the technical-equipment manuals. Also, since we are primarily interested in the editorial function, I shall not describe in detail the various types of technical-equipment manuals. To do so would require many volumes the size of this book. I shall merely touch on two points which have a bearing on the editorial process—their scope and cost.

Figure 13.1. A technical-equipment manual which could have been produced by a primitive technical writer.

Scope

For a weapons system, the basic types of technical-equipment manuals are: general manuals, equipment-oriented manuals, and depot-maintenance manuals. For one missile with which I am familiar, the information to be converted into manual form will make 20 sets of books, each of five volumes plus revisions—all told, 150 volumes whose print order will total 100,000 printed books. These are the texts for a 45-week key-personnel course on the operation and maintenance of a research-and-development prototype missile. Approximately 8,000 technical illustrations will have to be prepared in the process—1,400 schematic diagrams, 1,500 photographs and drawings, 800 graphs, 1,000 exploded views, 300 cutaways, 600 block and flow diagrams, and 2,500 tabular charts. It is obvious that such a program is a major financial undertaking.

Cost

The cost of technical-equipment manuals can best be understood by comparison with over-all equipment costs. The 1957 Armed-Forces missiles program called for an expenditure of well over two billion dollars—as much as was required to develop the atomic bomb. Multi-million-dollar contracts to individual companies are not uncommon. These high costs are causing a great deal of concern in many circles. At a late-1957 briefing, for example, worried top Air-Force officials told 100 representatives of 51 major contractors that future business would go where prices were lowest.

One of the ways of cutting down waste and inefficiency is to keep equipment operating properly. Complicated operating procedures must be clearly explained in the manuals which accompany the equipment. However, these manuals themselves are becoming more and more expensive. One reliable rule-of-thumb is that equipment manuals may cost 5 per cent of the cost of the equipment. According to its figures, the Department of Defense spent $250 million on publications in 1956. The product was a stack of publications 70 miles high. If the trend continues, it has been estimated that in 10 years the military's publication costs alone will reach perhaps one billion dollars. The cost of commercial publications is much more difficult to estimate. A study in this field would be a good subject for a Ph.D. thesis.

Such expenditures are justified only if the manuals are good and will perform the job for which they are intended. In the guided-missiles field this question has been asked, "Why do we need bookshelves of manuals on a specific missile if the missile is a one-shot job? If it performs its function it will be destroyed in the process. Why provide expensive manuals?"

The answer is that because the equipment is so expensive every last part must be checked thoroughly and economically in accordance with established procedures, so that the launching will be a success. And only the manuals can record the work of hundreds or thousands of scientists and engineers in convenient form.

To summarize, then, a technical manual is a written instruction guide for obtaining the most use and best service out of a product. Perhaps the shortest and most widely read "technical manual" is the one found on match-books: "Close cover before striking." The technical manuals I have worked with contain thousands of pages of text and illustrations, but the principle is always the same—the manual provides specific guidance and information for operating and maintaining the equipment, with precautions for minimizing damage to equipment and danger to personnel.

It is obvious that the preparation of complex, expensive, technical-equipment manuals involves many people. A hundred "publications" people or more may get into the act before the printed copies emerge from the mass of preliminary data. Where, then, does the editor fit in?

THE EDITOR AND THE EDITORIAL PROCESS

Many technical-equipment-manual projects, at least in the military field, take two, three, or more years to complete. The organization of the project, and the relation of the editor to it, is a major administrative problem which must be worked out for the particular project. Thus it is impossible to give a static organization chart which would serve as a general guide.

In some projects, the editor is responsible for the entire project, including the quality of the writing, artwork, typography, printing, and binding. In other projects his responsibility is limited to the checking of the manuscript. As in most fields, the amount of responsibility assigned depends upon the capabilities of the individual and the judgment of his superiors as to the degree of responsibility he can assume. We can ask this question: "What kind of person should the competent editor be, and what does he do to keep a major project moving?"

Like the old-maid school teacher, a stereotyped concept of a technical editor seems to be held by some, particularly nontechnical persons involved with other media. Based on this stereotype, the editor of a technical manual would be conceived of as a fatherly adviser who does not personally write technical manuals, but who instead spends his time encouraging others to write better manuals. Although inadequate, this generalized concept of the role of technical editor is a good point of departure for explaining the real function of the editor of technical-equipment manuals.

In the normal editorial process, there are two distinct functions which are performed. One is copy editing. The other is copy preparation for reproduction.

In the latter process, the editor is concerned with paragraphing, spelling, grammar, punctuation, references, retyping, marking for type, proofreading, and perhaps with layout, dummying, and other mechanical processes. This combination of functions is the one about which so much has been written and spoken, and from which the stereotyped concept has arisen. I contend that as far as the technical-equipment manual is concerned, these details can be performed by a competent assistant responsible to the editor.

I do not want the reader to get the impression that the preparation of copy for reproduction is unimportant—far from it. What I *do* want to get across is that in a complex project the editor must be capable of delegating responsibility, so that the entire project can be completed to meet tight deadlines.

Indeed, in dealing with very complex technical-equipment manuals the editor is the virtual captain of the intellectual ship. Let us summarize some of his major functions before we analyze them:

1. Defining the technical specifications.
2. Defining the style specifications.
3. Defining the nomenclature specifications.
4. Checking, checking, checking.
5. Directing the proofing program.
6. Personalizing the job.
7. Professionalizing the job.

Now let us analyze these functions.

Defining the Technical Specifications

Remember that a complex technical manual may cost much more than even the most expensive house. No one in his right mind would start building a house and then sit down to determine the specifications under which the construction will take place. Yet it is amazing how many writing projects, in the military field at any rate, are started just to "get the ball rolling." The editor is then called in to "polish up" the job. To complicate matters further, a major project may involve a dozen technical writers, each of whom has his own concept of the specifications. Most of these writers can prove that their interpretations of the specifications are correct by producing as evidence many volumes of acceptable publications.

What are the applicable technical specifications? For military technical-equipment manuals there are two types of specification—the general and the detailed. The general specification consists mostly of production information, such as final page size, type faces and sizes, general organization, and a liberal sprinkling of interpretations on English style, such as abbreviations and how illustrations shall be referenced. The detailed specification is, in effect, an outline of the proposed publication. It describes what material is to be covered and how it is to be organized.

For military equipment manuals there is a long list of several hundred such detailed specifications. The specifications for the missile manuals are summarized in Table 13.1. Even this long list of specifications, however, does not provide every answer. The more complicated the equipment, the greater are the chances that none of the applicable specifications will define the coverage of a particular technical-equipment manual. It is then up to the editor who will be charged with the responsibility of making the manual conform to the specifications to resolve all questions involved. This is often a long process, but as a patriotic citizen interested in economy, among other things, the technical editor must resolve these questions to make certain that each writer understands fully the nature of the monster he is creating.

For commercial equipment manuals the resolution of specifications questions is both more difficult and easier. It is more difficult because in most cases the project is unique for the company, and no specifications have been developed. It is easier because in the absence of rigid, predetermined specifications, the competent editor is relatively free to develop specifications which will produce

TABLE 13.1

TECHNICAL MANUAL SPECIFICATIONS AND EXHIBITS FOR ICBM* AND IRBM** CONTRACTORS

Number	Date	Amend.	Date	Title
GENERAL				
MIL-N-4350 (USAF)	10/1/5			Negatives for Reproduction, General Preparation of
MIL-H-4410A (USAF)	6/24/52	2	8/23/54	Handbooks and Illustrated Parts Breakdowns: Title Pages, "A" Pages and Reproduction Assembly Sheets: General Preparation of
MIL-STD-218-1, -2, -3	7/1/55			General Preparation of Technical Manuals
MIL-H-5474A	9/11/50	1		Handbooks and Breakdowns: General Preparation of
MIL-N-7384A (USAF)	11/8/55	1	2/13/56	Notices: Contractor-Furnished Aeronautical Equipment and Accessories
GUIDED MISSILES				
*Exhibit WDT-56-3A***	7/1/56	1	10/1/56	Exhibit for: Manuals, Technical, Organizational-Level, General and Job Oriented Operation-Maintenance for Ballistic Missile Weapons Systems WS107A-1, WS107A-2, WS315A
MIL-H-9210B	12/20/55			Handbook: Inspection Requirements
MIL-H-8031A (USAF)	5/3/55	1	8/18/55	Handbook: List of Applicable Publications
MIL-B-5005A	8/17/51	1	10/16/52	Breakdown: Illustrated Parts
MIL-T-7578D (USAF)	11/18/55			Time Compliance Technical Orders

* Intercontinental Ballistics Missiles (ICBM).

** Intermediate Range Ballistics Missiles (IRBM).

*** New Exhibits and Standards for WS107A-1, 107A-2, 315A Associated Contractors.

TABLE 13.1 (Contd.)

Number	*Date*	*Amend.*	*Date*	*Title*
AERONAUTICAL ENGINES				
Guided Missile Engines				
MIL-H-25394 (USAF)	9/19/55			Handbook: Field Maintenance Instructions
MIL-H-7985	11/5/52			Handbook: Overhaul Instructions (Liquid Propellant Rocket Engines)
MIL-O-8029 (USAF)	9/16/52			Overhaul Change Technical Orders
MIL-B-5005A	8/17/51	1	10/16/52	Breakdown: Illustrated Parts
MIL-T-7578D (USAF)	11/18/55			Time Compliance Technical Orders
ACCESSORIES				
MIL-H-25393 (USAF)	9/20/55			Handbook: Field Maintenance Instructions
MIL-H-25095 (USAF)	2/17/55			Handbook: Field Maintenance Instructions
MIL-H-6813	8/3/50	2	12/8/55	Handbook: Overhaul Instructions
MIL-H-6738A	7/25/52	1	12/23/54	Handbook: Overhaul Instructions with Parts Breakdown
MIL-B-5005A	8/17/51	1	10/16/52	Breakdown: Illustrated Parts
MIL-T-7578D (USAF)	11/18/55			Time Compliance Technical Orders
AIRBORNE ELECTRONICS				
MIL-H-25095 (USAF)	2/17/55			Handbook: Field Maintenance Instructions
MIL-H-6814A (ASG)	3/6/53			Handbook: Overhaul Instructions
MIL-B-5005A	8/17/51	1	10/16/52	Breakdown: Illustrated Parts
MIL-T-7578D (USAF)	11/18/55			Time Compliance Technical Orders

TABLE 13.1 (Contd.)

Number	Date	Amend.	Date	Title
SPECIAL WEAPONS				
MIL-H-9128A (USAF)	9/27/55			Handbook: Special Weapons Procedures
MCMTTP Exhibit No. 17	10/25/55	1	12/23/55	Handbook: Assembly, Service, Storage and Surveillance Instructions
GROUND ELECTRONICS				
MIL-H-6757A	3/6/53	1	8/23/55	Handbook: Field Maintenance Instructions
MIL-H-4963 (USAF)	2/15/55			Handbook: Preventive Maintenance Instructions
MIL-H-6814A (ASG)	3/6/53			Handbook: Overhaul Instructions
MIL-H-6544A (USAF)	7/26/55			Handbook: Installation Instructions
MIL-H-4978 (USAF)	5/18/55			Handbook: Standard Installation Instructions
MIL-B-5005A	8/17/51	1	10/16/52	Breakdown: Illustrated Parts
MIL-T-7578D (USAF)	11/18/55			Time Compliance Technical Orders
GROUND SERVICING AND HANDLING EQUIPMENT				
MIL-H-7960A (ASG)	2/5/53			Handbook: Operation and Service Instructions
MIL-H-7991	10/21/52			Handbook: Overhaul Instructions
MIL-H-6261A	12/18/52			Handbook: Operation, Service, and Repair Instructions (Condensed)
MIL-B-5005A	8/17/51	1	10/16/52	Breakdown: Illustrated Parts
MIL-T-7578D (USAF)	11/18/55			Time Compliance Technical Orders
TRAINING DEVICES				
MIL-H-4885 (USAF)	4/15/55			Handbook: Operations and Maintenance Instructions

TABLE 13.1 (Contd.)

Number	*Date*	*Amend.*	*Date*	*Title*
TRAINING DEVICES (*Contd.*)				
MIL-H-8074A (USAF)	2/13/56			Handbook: Inspection Requirements
MIL-H-8031A (USAF)	5/3/55	1	8/18/55	Handbook: List of Applicable Publications
MIL-B-5005A	8/17/51	1	10/16/52	Breakdown: Illustrated Parts
MIL-T-7578D (USAF)	11/18/55			Time Compliance Technical Orders
NON-AERONAUTICAL ENGINES				
MIL-H-4884 (USAF)	10/21/54			Handbook: Overhaul Instructions
MIL-B-5005A	8/17/51	1	10/16/52	Breakdown: Illustrated Parts
MIL-T-7578D (USAF)	11/18/55			Time Compliance Technical Orders
AUTOMOTIVE				
MIL-O-26788 (USAF)	6/28/55			Handbook: Operation and Service Instructions
MIL-O-26787 (USAF)	6/28/55			Handbook: Overhaul Instructions
MIL-B-5005A	8/17/51	1	10/16/52	Breakdown: Illustrated Parts
MIL-T-7578D (USAF)	11/18/55			Time Compliance Technical Orders

an economical manual with complete technical coverage. A common solution is for the editor to modify existing military or other government specifications to meet his particular requirements. This is often a good solution because, in spite of criticisms which I have heard, government specifications are carefully developed. It may take years to develop one specification, and the approved specification represents the best current thinking.

In the guided missiles field, for example, the specifications problems have been found to be so unique that special specifications are prepared for each weapons system. For one weapons system, the specification itself runs to 76 pages.

Defining the Style Specifications

In addition to technical specifications, there are many style specifications and guides. Attempts to standardize these guides have only resulted in more confusion. For every major project it is the duty of the editor to provide each writer with a full set of style specifications which will be applicable to the particular project, consistent with the requirements of the technical specifications. These should be explained by simply stating that these style specifications are the ones to be used for the particular project. Individual preferences have to be submerged for the common good.

Defining the Nomenclature Specifications

As products have become more complex, it has become increasingly difficult to describe them. New words have been created, and old words have been ornamented with descriptive adjectives. To take a very simple example, a bill of material calls out an air-drying, fungus-and-moisture-resistant varnish. It is obvious that if this varnish must be referred to 10 or 20 times the complete nomenclature would make for pretty heavy reading. The editor should develop a consistent shortened nomenclature which should be followed by all writers.

Checking, Checking, Checking

Merely specifying what is to be done is not sufficient. It is the editor's job to check each writer's work, time and time again, especially early in the project. This will save time and money in the long run.

Directing the Proofing Program

A technical manual which is not wholly accurate is worthless. More important, it is dangerous. It is dangerous from the point of view of the company's reputation. It is dangerous because inaccurate procedures may result in damage to the equipment and loss of life or limb.

The best way to check out a technical manual is by means of a proofing program, which involves actually checking out the procedures against the work being performed. This may be done by the technical writer during the course of the project. However, the writer is under heavy pressure to complete a project and may find it difficult or impossible to do a thorough job. It thus becomes the responsibility of the editor of a technical-equipment manual to proof out the manual.

The project may be so complex that several editorial assistants may be needed. Can editorial assistants do the job? The general philosophy back of most procedural manuals is that they must be able to be understood and followed by a man who has no more than a high-school education. Thus, if a manual cannot be followed by an intelligent editorial assistant, it will probably be of little value to the man in the field and will have to be rewritten to make it useful.

Personalizing the Editorial Job

It is in this area that I sometimes encounter disagreement with my point of view. By personalizing the editorial function I do not merely mean that a good working rapport must be established between the editor and the technical writers. I mean that the editor should, if possible, be personally capable of producing a manual or a section of a manual when the need arises, if only for the purpose of understanding for himself the problems which writers encounter.

When I made this statement a number of times at various technical-writing-society meetings, I found that the reaction from some was rather violent. However, most of the disagreement came from editors whose technical backgrounds, I found upon inquiry, were nonexistent or extremely limited. Remember that in this chapter I am referring to complex technical-equipment manuals. Their writers are graduate electronics engineers or men with equivalent training and experience. These men will usually take criticism or accept suggestions only from one of their peers.

I do not mean to imply that a man with a good conventional background cannot grow into the job. One of our best editors has a Ph.D. in English literature. However, this man was not lazy. He found that to hold and advance in a job with us he needed a stronger technical background. He went back to school, studied electronics, and is now confident that he has a solid understanding of his job. He commands the respect of the entire staff. He is a member of the team, not its mascot.

If only as an intellectual exercise, then, the ideal editor should write on occasion.

Professionalizing the Editorial Job

In the technical-writing field, the word "profession" has been used very frequently. Most of the time there is the hopeful wish that technical writing and editing will grow into a profession. But what constitutes a profession is often not clear. The best article on the subject that I know of is the one by Vannevar Bush which appeared in *Science** in January 1957. It would be difficult to summarize this clear presentation adequately, but I would like to call your attention to one short statement:

"Members of a profession . . . do not merely advise, they insist upon being heard; they do not subject their opinions for the judgment of the layman who is their client, no matter how powerful he may be; they insist that they have his confidence and that, in their special field, this opinion shall control, or that the client turn elsewhere."

I know that it is difficult in this practical world to achieve this goal, but if we can get close to it, we can make technical editing a respected profession.

SUMMARY

Before really summarizing, I must discuss one extremely important point which deeply concerns me. That is the relationship between the quality editing of equipment manuals and their cost. Quality editing is time-consuming, but it makes the difference between a good and a poor job. If the cost of a project is estimated on a competitive basis, it is, I contend, almost impossible to allow for a quality editing job on a major project and also bid competitively. In the rough-and-tumble competitive estimating procedure, it is the editing which suffers most, because this is an intangible which cannot be pinned down. I would like to suggest that the entire quality-control process be exempt from competitive bidding and negotiated as a separate item after a rough draft has been approved. Then, based on the size of the project, the amount of editing can be worked out. If a prospective client says to me: "I don't care anything

* Vannevar Bush, "Professional Collaboration," *Science,* **125,** 49-54 (1957).

about quality—just give me a book which will satisfy my contractual obligations," as a professional man I prefer that the client place his teratological literature requirements elsewhere.

To summarize then, a good editor of a complex technical-equipment manual cannot withdraw into his ivory tower. He must actively pursue the project from its inception. He cannot permit himself the luxury of sitting behind his desk and awaiting the flood of material which will come to him from many very individualistic writers. He should learn as much as possible about the technical aspects of the project. The best informed editor does the best job.

Perhaps the best way of defining the functions of a person who can assume all these responsibilities is in terms of a "quality-control literature inspector." It is this person's influence that will convert rough, unappealing manuscripts into fine-looking, smooth-reading publications which will keep our country technically strong in war and peace.

Part IV

EDITING GRAPHIC AIDS AND OTHER EXHIBITS

Chapter Fourteen

Editing Slides and Other Graphic Aids

DAVID N. COLLINS
Administrative Assistant, Chemical Research
Chrysler Corporation
Detroit, Michigan

and

B. A. JONES
Supervisor of Services, Chemical Research Division
Ethyl Corporation Research Laboratories
Detroit, Michigan

The traditional act of presenting a technical paper consists of one scientist, talking. The materials he uses to present his data and to heighten its impact on the listener make up the "graphic-aids package." Serious consideration of these materials apart from their technical content may be new to scientists and engineers, but would not be strange to educators or advertisers. Our necessity for using the purely advertising jargon word "package" gives an idea of the importance of these materials to the advertiser; the scientist must borrow a word to describe them. This chapter will present some readily available techniques for increasing the usefulness of graphic aids in a technical presentation.

THE FOUR FUNCTIONARIES

Some authors prepare their own graphic aids, others can call on a highly skilled Presentations Department, but all graphic aids are the product of four separate functionaries—the author, the editor, the artist, and the artisan. The author decides what to say. The editor knows how to say it. The artist puts the plans on paper, and the artisan (photographer, carpenter, draftsman, or dressmaker) builds the finished product.

Frequently, two of these functions are combined in one person, such as artist-artisan, or editor-artist. However, in any large organization, efficiency requires that the editor stick to editing, and the photographer to photography. Most important, no author should be his own editor, for the improvements of editing come as much from the fusion of two independent viewpoints as from any special "editorial" skill.

As the middleman between science and the arts, the editor is best able to set and maintain standards of quality in graphic aids. Most authors lack knowledge of the artist's techniques; the artist can give little attention to each author's technical problems. The editor earns his keep by having a foot in each field.

The Author Is Boss

If there is such a thing as a supervisor in this four-man team, it is the author. He is the authority on the subject to be presented, and will make use of the results. The editor has only a specialist's advantage: a greater knowledge of display techniques and of their production. Using this knowledge, the editor guides and suggests until a program agreeable to the author is established. (With only the specialist's advantage, the editor has also the specialist's recourse: one real "clinker" demanded by the author and used over the editor's objections will produce more future cooperation than a dozen unsupported arguments.)

A practicing editor, even a part-time one, achieves his desired level of quality by establishing a give-and-take relationship with the artists who prepare his layouts and the artisans who process them. Here, the editor is on the other side of the specialist's fence. He may be skilled enough to design or sketch a layout, and does approve the artist's execution; but good ideas in display must combine the technical knowledge of the editor with the artist's knowledge of his art. Neither alone is enough.

The editor is also the contact between author and artist. This is not only for the author's convenience, but also for protection of the level of quality. Unless the author's authority is filtered through the editor's skill, it is easy for the author to summarily overrule every suggestion of the artist, no matter how well founded artistically.

Routine Photography

Beyond the artist's level the last purely scientific consideration disappears, leaving the photographer or other artisan free to do a routine (but routinely good) job on the instructions and layouts furnished him. This concentration of artistic skill in the hands of the artist is most important in a large organization; the photographer has little time to spare for advice on projects which occupy only a small fraction of his work load.

In summary, the editor is the prime mover of the team which prepares graphic aids. He represents the artist to the author, and the author to the artist, to the advantage of both.

THE EDITOR LEARNS

The genesis of a graphic-aids package is a conference between editor and author. The editor must learn from the author the details of how the presentation will be made. Who is the audience? How many people, and why are they assembled? Where is the meeting room; what size and shape? Is there projection equipment? A stage? Finally, what is the purpose of this presentation? To inform? To sell? To entertain? The answers to these questions outline the editor's problems and possibilities.

THE EDITOR CONSIDERS

Audience Requirements

Graphic aids come in many shapes and sizes, to fill many different requirements. A rough classification by type would include models and test pieces, data charts, pictorial sequences (movie or still photography), associative symbols, and key-word outlines. Not every type can be used with every size of audience. Models and test pieces are best for table-top display, limited to audiences of five or six persons. Art-work charts, blackboards, and flannelboards do very well as wall displays for a maximum audience of 20. No more than 200 can view slides or movies in the usual meeting room; above this size remain only theater screens, multiple meetings, and television.

Television as a graphic aid is worth an extra word. It is glamorous and expensive, and has most commonly been used for the advertising value of its novelty. Above this real, though intangible, value is a less obvious characteristic. It is the only medium except printed matter which reaches a large audience assembled in groups of three or four. In effect, this is table-top and blackboard lecturing, multiplied by the number of television receivers available.

Slides

Curbing his fancy to fit his budget of time and money, the editor usually decides on a slide presentation. Except for very small groups, or for a perambulating audience, this is a wise choice. Slides are easy to prepare, inexpensive, and permanent. They give visual impact in a wide variety of effects. On the other hand, they are rigidly limited in length of copy, and, except for a few color effects, are completely two-dimensional.

Since slides are so widely used, it is odd that their many effects are not more completely exploited. Slides are not technical illustrations; most good slides look enthusiastic and rather simple-minded in print, and few journal illustrations ever made good slides. The purpose of a slide is the single-minded presentation of a single fact or facet of data, to accompany no more than 60 seconds of the spoken word. Journal illustrations may be more complex; they are meant for repeated reference and study. Good slides are made from illustrations by redrawing to slide proportions—legends are made bolder, contrasts heightened, details simplified. If the illustration demands lengthy analysis, make several slides. Compromises here are always in the direction of mediocrity.

Positive Slides

The usual positive slide (black lines on a clear background) offers several difficulties in preparation and use. Two photographic steps are required from the original art work: a film negative, followed by a glass or film positive.

If it is on glass, the positive is easily cracked or broken, and its projection is marred by dust specks and fingerprints on the clear areas. The slide cannot be altered or repaired; such changes must be made on the negative, and a new slide then printed. A much more flexible medium is the negative slide, which is the film negative itself, bound between two sheets of clear glass.

Negative Slides

The negative slide, of course, projects white lines on a black field. Its preparation requires only one photographic step, after which it can be spliced, patched, colored with transparent water colors, or opaqued. Figure 14.1 is an artist's layout, drawn for a negative slide employing three colored filters. Chart lines, grids, letters, or whole words can be added or deleted, using nothing more than film clips, a razor blade, and black tape.

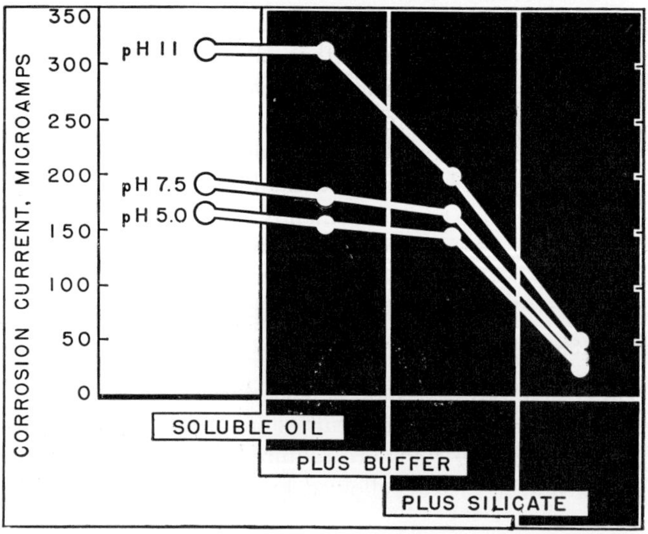

Figure 14.1. Artist's layout for a three-color negative slide.

Even journal illustrations can serve as originals for negative slides. The bottom half of Figure 14.2 has been lined over for making a negative slide. A new layout would have been better, for bolder lettering could be used. Since the points of interest (lines, letters, etc.) are also those of greatest light intensity, negative slides project well in a half-lighted room, or even in full room light.

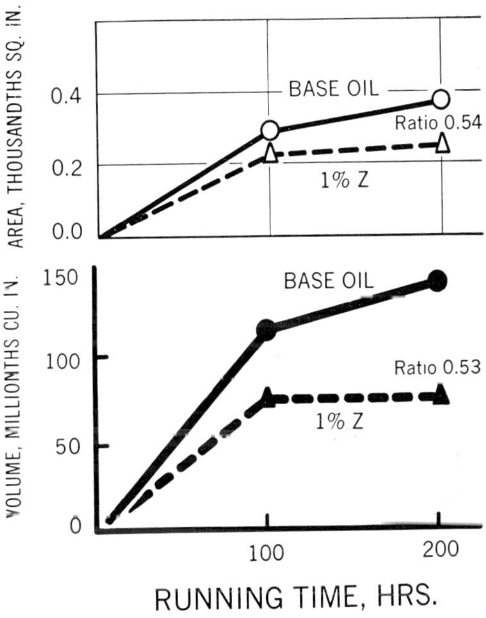

Figure 14.2. Bottom half shows how a journal illustration (*top*) may be converted to a slide original.

Color Slides

Color photography is also valuable in using slides, either to record colors which are technically important, or as a change from the pen-and-ink art work of the draftsman. In color slides, the background is as important as the subject matter, and as varied; these slides are made from table-top arrangements of models and cut-outs, like still-life photography.

Slides as Poster Art

As the slide collection has become more important to the technical presentation, the slides themselves have changed to a less formal, more emphatic medium. From rigid respectability, slides have become a form of poster art—bright colors and bold lettering to catch the eye, color accents to make a technical point, exaggerated shapes and dimensions. The same art work that sells nylons will also elucidate "reactivity differences between diastereomers in the Wagner-Meerwein rearrangement," if required.

As slide composition becomes more varied, it becomes possible to use visual impact to emphasize each point of the technical presentation as it appears. Visual impact is greater than aural; a well-prepared slide will out-shout the most leather-lunged lecturer. This technique of slides in continuity, as opposed to slides as illustrations, calls for 20 or 30 slides for a 20-minute presentation—a not unreasonable number if slide preparation is the routine province of a Presentations Department. The routines of such preparation have been described in detail by publications of the Eastman Kodak Company (built around color photography, of course).

THE EDITOR EDITS

Every technical presentation brings its own graphic-aids problems; however, most of the editor's job consists of the routine planning and preparation of slides. A timetable of operations is essential, both to help the author's preparation of his talk and to prevent unreasonable demands on artist and photographer. The following schedule for lantern-slide preparation is a practical compromise between laissez faire and a frantic rush.

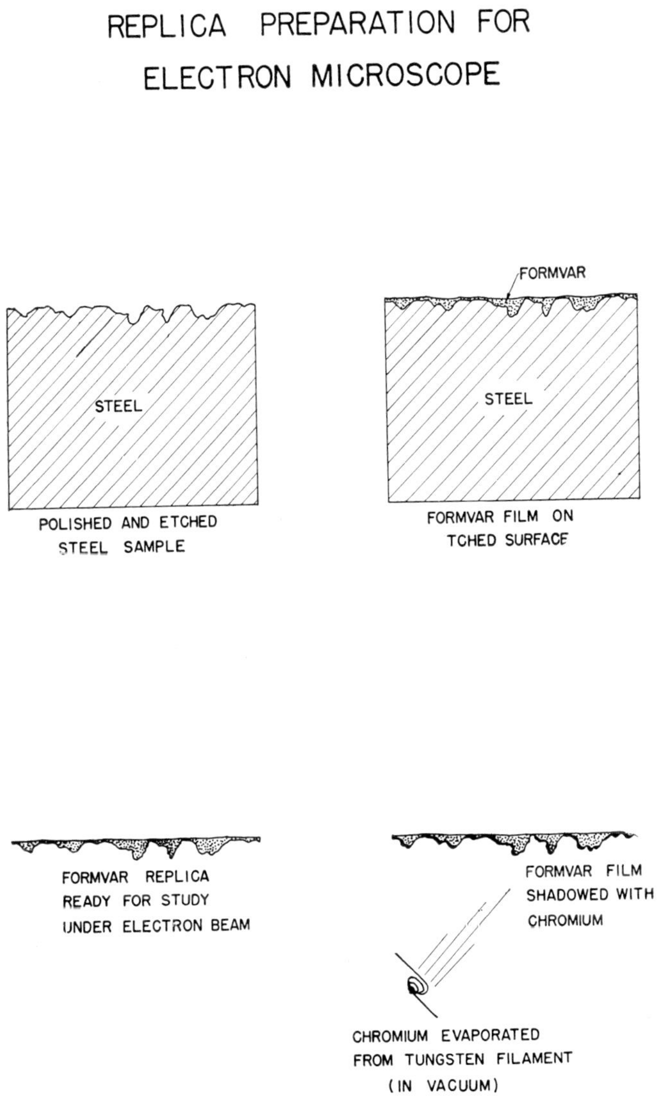

Figure 14.3. Wall chart, unsuitable for a slide original.

Author's Conference

One month before the slides are due, the editor has his conference with the author, at which the style and number of illustrations are selected. Immediately, the editor releases his sketches to the artist for preparation of the layouts. Two weeks before the due date, the art work must be sent to the photographer. This allows a week after photography for coloring negatives, binding the slides, and making last-minute corrections.

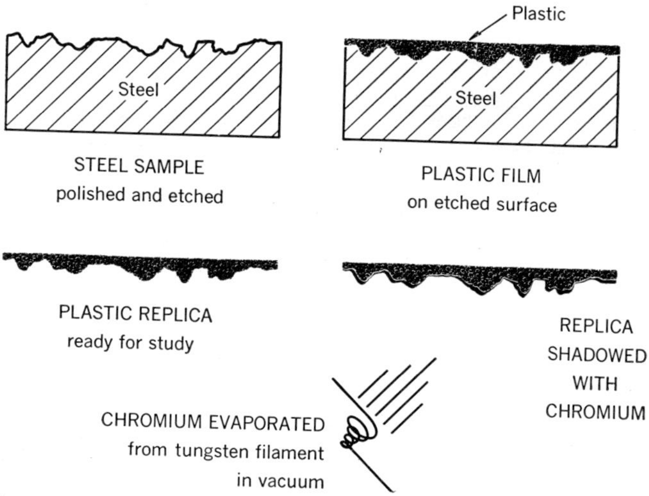

Figure 14.4. Negative-slide layout based on Figure 14.3.

The graphic-aids editor differs little from any technical editor in his attention to style and technical content. However, his province contains less "What?" and more "How?" in the technical presentation, and he must withstand even more of "Why not?"

The Word is NO!

An editor of slides must learn to say "No!" This is prescribed by the rigid limits of his media. A journal editor may pass superfluous detail in an illustration, or an extra sentence in a paragraph, with few penalties. On a slide, however, even one extra line of type makes every word harder to read, and extra detail in a slide drawing may make it completely incomprehensible.

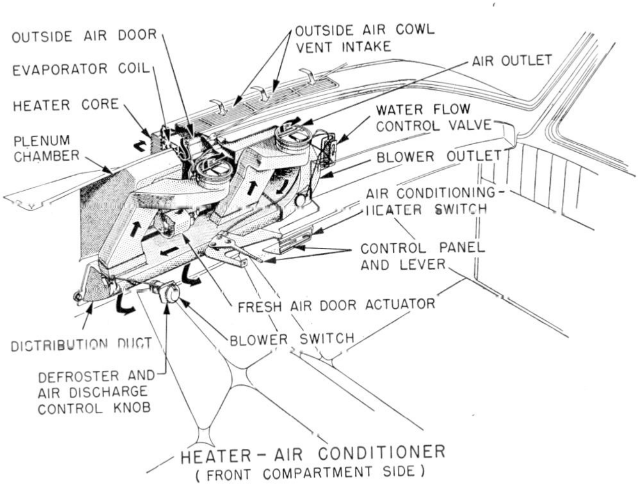

Figure 14.5. While this drawing is excellent for extended study, it is far too complicated to make a good slide.

After the firm "No!", however, the editor must be ready to condense and rephrase, to split tables and to crop drawings, until all the author's material has best been presented. Authors are more important than editors, and the final compromises usually reflect this. But it is very rare copy which cannot be condensed by concentrated effort. Titles may be dropped, legends condensed, curves omitted. Skilled surgery is usually beneficial.

Examples

Editing of the text and a new layout were required to transform the wall chart of Figure 14.3 into a negative slide layout (Figure 14.4).

RELATIVE COSTS OF SHEET METAL

(Black Iron, 1 Sq. Ft. 0.048" Thick = 100)

	THICKNESS (INCHES)	COST PER POUND	COST PER SQUARE FOOT
Black Iron	0.048	50	100
Galvanized Iron	0.030	65	95
Terne Plate	0.030	70	90
Aluminum 3SH-14	0.032	275	125
Stainless Steel 302	0.025	350	370
316	0.025	485	510
Copper	0.032	355	530
Monel	0.031	610	855

TYPICAL
RESIN-BONDED SAMPLE

	VOLUME PERCENT	WEIGHT PERCENT
TEFLON	20	21
RESIN	30	17
FILLER	50	62

Figure 14.6. This detailed table contains too many entries to make a good slide.

A good technical illustration for extended study is Figure 14.5. As a slide, however, it is hopelessly complicated. Individual parts of the drawing should be presented in a number of slides to fit the presentation as it develops.

APPLICATIONS OF CORROSION TESTING

1. STUDY CORROSION MECHANISMS
2. SELECT MOST SUITABLE MATERIAL FOR A GIVEN ENVIRONMENT
3. DETERMINE CORROSION RESISTANCE OF A SPECIFIC MATERIAL
4. DEVELOP NEW MATERIALS
5. CONTROL PRODUCT UNIFORMITY

WHY CORROSION TESTING

1. Study corrosion mechanisms.
2. Select resistant materials.
3. Measure corrosion resistance.
4. Develop new materials.
5. Control product uniformity.

Figure 14.7. Omitting even a few words improves a slide. At top, the author's version; at bottom, the final copy.

Detailed tables like the first one in Figure 14.6 contain too many entries to fit a single slide. This one is better presented graphically. If only a few entries will be discussed, these should be abstracted, as the second has been.

Even a few words omitted mean a more legible slide. At the top of Figure 14.7 is the author's version of an outline; at bottom, the final product.

THE EDITOR ADVISES

Distracting Details

Not all authors are skilled in lecturing, and the editor may make himself useful by recommending details in the presentation which are not strictly within his province. Outside distractions break communication between the author and his audience. Flicking the lights on and off between scattered slides, stray shadows from the lecturer's head or arms on the projection screen, flurries of activity around the slide projector—all these distract the weary listener at the expense of the subject matter. There should be no breaks in a slide sequence; the main topics in the subject outline benefit even more from screen presentation than do the more usual illustrations.

Is It Funny?

Humor is a spice. It should never monopolize the visual part of a technical paper.

Slides as Notes

If the details of a paper are well covered by slide continuity, the slides themselves may well serve as the author's notes. Contact prints of the slide negatives can easily be arranged in order, eight to ten to the page. Such notes, beside their covering of the subject, permit the lecturer to move easily from slide to slide, withholding the appearance of the new slide until the moment of maximum impact.

INDIVIDUALITY—THE STEP BEYOND QUALITY

This may well be the first document on technical slides which does not recommend adherence to uniform standards for slide preparation. The American Standard Y-15.1, "Illustrations for Publication and Projection," issued in 1958, will certainly "provide a firm basis for revision of the generally outmoded pamphlets now issued by various scientific societies." But this is only the minimum, the point from which a good editor starts. Beyond this standard lies the whole range of commercial art. Individual artistry and experiment will produce the best graphic aids.

ACKNOWLEDGMENTS

While the opinions expressed in this paper are those of the authors, few of the devices employed are original. The separation of editorial functions from those of photography was well presented by Norman Salmons, speaking for Eastman Kodak Company, at a meeting of the Michigan Section of the Industrial Photographers Association. Negative slides in color have been used for many years at Ethyl Corporation Research Laboratories. The slide masters used here as illustrations were selected from articles edited by the senior author at Chrysler since 1955. They are the work of Garnet K. Ross of Chrysler Engineering's Research Section, who also helped work out many technical problems in editing. Valuable assistance has also been given by the Photographic Section and the Presentations Department of Chrysler Engineering.

BIBLIOGRAPHY

1. American Standards Association, "Illustrations for Publication and Projection," New York, American Standard Y-15.1, 1958.

2. Connelly, W. J., "Mechanical Aids to the Effective Presentation of Technical Papers," Paper presented before the Division of Chemical Literature, 128th Meeting Am. Chem. Soc., Minneapolis, 1955.

3. Jones, B. A., "Slides: Confusing or Clear?", Paper presented before the Division of Chemical Literature, 122nd Meeting Am. Chem. Soc., Atlantic City, 1952.

4. Staff, "Photographic Production of Slides and Film Strips," Rochester, N. Y., Eastman Kodak Company.

Chapter Fifteen

Editing Illustrations for Technical Reports and Papers

ETHALINE CORTELYOU
Literary Analyst, Armour Research Foundation
Chicago, Illinois

and

B. A. JONES
Supervisor of Services, Chemical Research Division
Ethyl Corporation Research Laboratories
Detroit, Michigan

INTRODUCTION

The predominant characteristic that distinguishes technical writing from other expository writing is the reporting of technical data. Scientists specialize in the collection, interpretation, reporting, and use of technical data, which may be presented in detail in tables. Usually, however, numerical relationships of variables are of more interest than measured or calculated absolute values; such relationships are often presented as equations, graphs, and bar diagrams.

An examination of reports from many research laboratories indicates that, typically, the illustrations and tabular material are less carefully edited than the text. They are less consistent in form, and more likely to contain errors in spelling and grammar. A possible explanation is that even the least experienced technical editor will edit text material with some degree of confidence because he claims a proficiency with words; moreover, he does not hesitate to make word changes because the author's draft can seldom be used as such for a duplication master, and must be retyped or typeset.

Although tables also may have to be retyped or typeset, they often are edited only for spelling and for consistency of format because it is assumed that the person who collects the data is the best qualified to determine how they should be reported. Many editors do not edit equations because they do not feel capable of doing so. Errors are less likely to occur in equations, however, because a technical author's mathematical competence is usually greater than his literary skill. Furthermore, his supervisor is expected to check equations carefully during his technical review of the report or paper.

THE IMPORTANCE OF PROPER SCHEDULING

In technical-report writing, a picture may or may not be worth a thousand words, but it certainly costs more to reproduce than the words it replaces. An author seldom does his own photographic work, except in such specialties as microscopy or spectrography. Even then, he usually produces only the originals and has copies made by photographic services. Although an author may plot rough graphs of his data, he usually has them copied by draftsmen. Since both photography and drafting are time consuming, such work must be ordered well in advance of the due date for the first draft of a report, manual, or paper.

Often, in ordering titles and labeling for his illustrations, an author may be as careless as he is in writing the first draft of the report. Errors such as misspellings or the use of improper abbreviations are more easily corrected in the text two weeks before the mailing date of a report than in a graph or in an expensive glossy photograph (sometimes, unfortunately, as many as 40 copies of it) two days before the mailing date, which may be the first time that the editor sees the illustrations an author has ordered. Also, if errors do escape detection and appear in the printed copy, they are less obvious in a page of text than on a sparsely worded illustration.

Schedule-conscious journal editors avoid such problems by printing carefully detailed directions to authors for the preparation of illustrations. Because a journal editor can demand good illustrations from an author at the author's expense, he has no compunction about making necessary changes. A report editor, however, may be tempted to use unsuitable illustrations because considerable money may have already been expended in their preparation. For this reason, a report editor's instructions to authors on the preparation of illustrations should be as detailed as possible.

EDITORIAL RESPONSIBILITY AND CONTROL

Most technical editors are of the opinion that the ideal report processing setup would give the technical editor control over report illustrations, either by placing photographic and drafting services under his control, or by requiring the editor's signature on any order for drafting or photographic work on report illustrations. For example, the department at Armour Research Foundation which consistently produces those reports containing the most uniform and best illustrations is the department in which the editorial unit has its own draftsmen. Chapter Two describes a similar arrangement.

Good results are usually obtained from a report-processing setup in which a graphic-arts group prepares all report illustrations—photographs, graphs, line drawings, and sketches. At Battelle Memorial Institute, for example, one group has the over-all responsibility for all mechanical details of processing a report—typing and duplicating as well as producing illustrations. This group works closely with the technical-editing group to process reports on schedule. In effect, this assigns to the report-processing group the responsibility for the attractiveness and legibility of a report. Such a setup should result in particularly good illustrations because the advantages and limitations of all duplicating processes available should be considered in deciding what type of illustration masters should be prepared.

STANDARDIZED INSTRUCTIONS

A good illustration guide is an important factor in obtaining uniform illustrations whether the graphic-arts service is within the editor's own group or is an outside group. Because of the time and cost elements involved, there is ample justification for having the section that deals with illustrations in a report-preparation manual much more detailed and extensive than the treatment of text. An important function of any report-preparation manual is the establishment of routines, so that the need for making decisions on the mechanical details of processing a report is kept to a minimum. When the need for minor decisions is avoided, better author-editor relationships result, for then the chances of disagreement are reduced. Actually, author-editor relations should be such that an author routinely discusses his illustrations with the editor before ordering photographic or drafting work, since there is no pride of authorship involved. The author should be encouraged to consider such consultation as a routine service that he can expect from an editor; if the editorial service is efficient and practical, he will demand it as a service to relieve him of petty details in writing his report.

The illustration guide should have definite standards for such details as the types and sizes of illustrations, the weights of lines and captions in drawings and photographs, the background in photographs, and the use of upper and lower case letters in titles and labels.

A published standard such as "Illustrations for Publication and Projection," Y15.1-1958 American Standard may be adopted, or a special guide may be tailored to meet the needs and reproduction facilities of a particular organization. Such a guide should represent the consensus of the technical editor and the heads of the drafting, photographic, and duplicating services. The references listed in the bibliography of this chapter will be of assistance in developing a guide; the manuals of such organizations as Battelle Memorial Institute and Knolls Atomic Power Laboratory will also be useful.

THE VALUE OF EXAMPLES

The presentation of examples in the illustration guide is a very useful device. A line drawing and a photograph of the same piece of equipment, the same assembly, or the

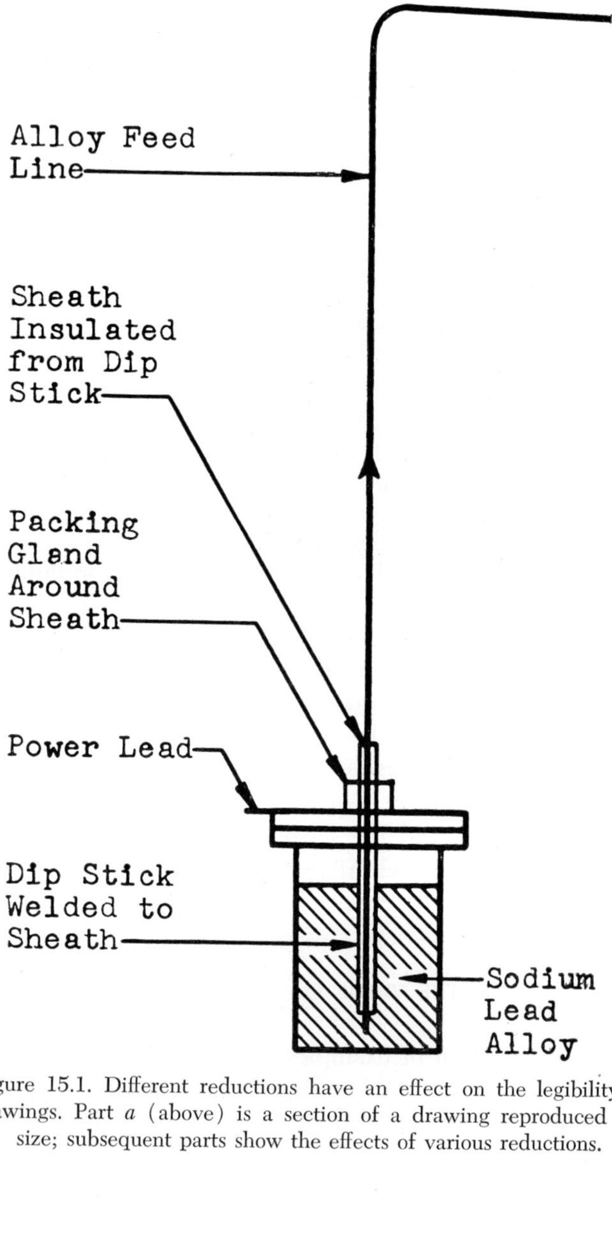

Figure 15.1. Different reductions have an effect on the legibility of drawings. Part *a* (above) is a section of a drawing reproduced full size; subsequent parts show the effects of various reductions.

same instrument may help an author to decide whether to use a drawing or a photograph for a specific illustration. Comparison of the same photograph printed on glossy paper, on Ad-Type paper, and as an offset half-tone may be quite instructive for an author who usually knows little about duplication processes. If a rough estimate of prices and times involved are included, an author may decide in

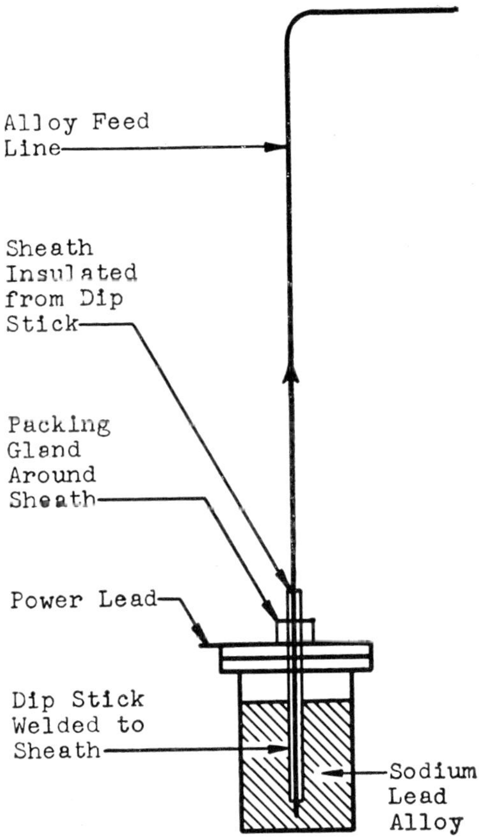

Figure 15.1-*b*. Same section shown in Part *a*, reduced 20%.

favor of offset half-tones even for micrographs, provided sharp detail is not needed.

Examples of different reductions of the same illustration (Figure 15.1) would be a most worthwhile addition to any illustration guide, because few persons can visualize the size of final copy when reduction is involved. For some reason, many authors are unable to appreciate that a reduction not only reduces line width but also reduces the area of white, or unused, space surrounding it. Also, many find it difficult to understand the differences in legibility obtained by a one-fourth reduction and by a one-third reduction.

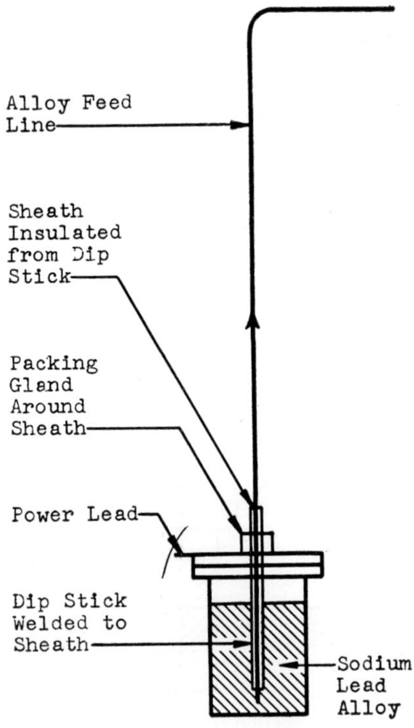

Figure 15.1-*c*. Same section of copy in Part *a*, reduced 33%.

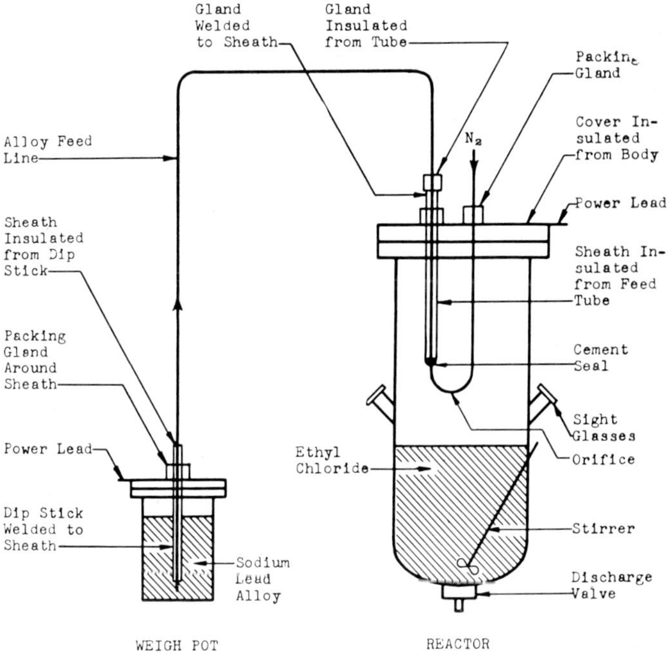

Figure 15.1-*d*. Copy reduced 50%.

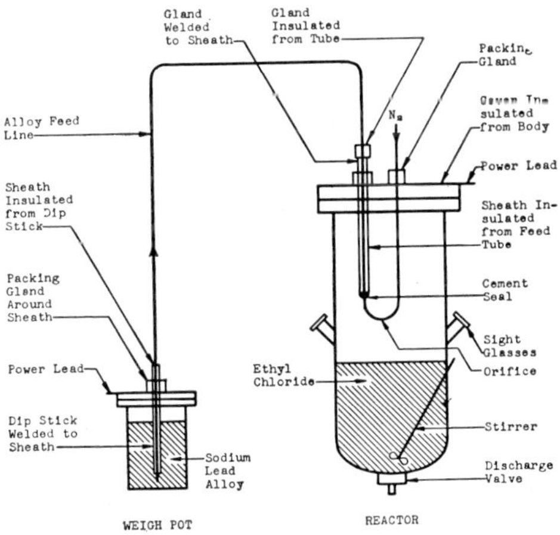

Figure 15.1-*e*. Copy reduced 65%.

Examples (Figure 15.2) of how a photograph of equipment might be improved by cropping, by enlarging only a certain area, or by opaquing unnecessary details should be offered.

Still other examples might be used to illustrate the un-

Figure 15.2-*a*. As may be seen in Part *b* (next page), clarity and emphasis are gained by cropping and enlarging the significant area in the photograph shown above.

desirability of reproducing the very fine grid lines in a graph by showing the same data plotted on a fine grid and on a graph in which only the major coordinates were reproduced, or on a graph on which the major coordinates are merely indicated by tick marks at the axes of the graph (Figure 15.3). Of course, it could also be pointed out that fine grids are unnecessary unless a graph is to be used for interpolation of data or for calibration purposes.

Figure 15.2-*b*.

It is important that an author understands that an illustration should be used only when it is needed to improve a report, and that both the budget and professional ethics dictate that illustrations should never be used to pad a report or manual. There is seldom any justification for

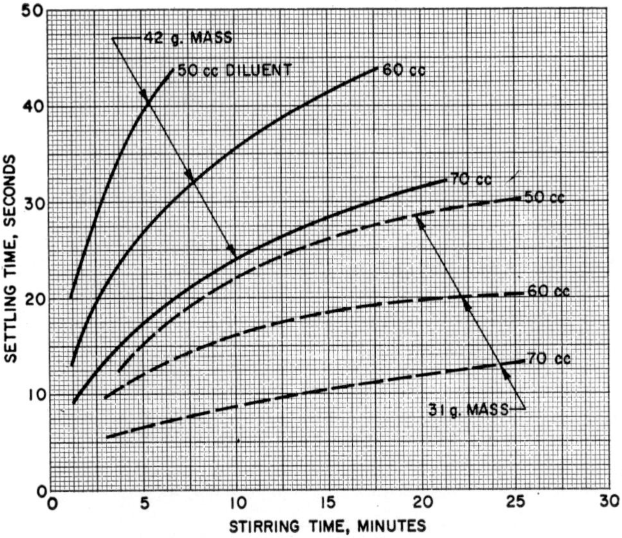

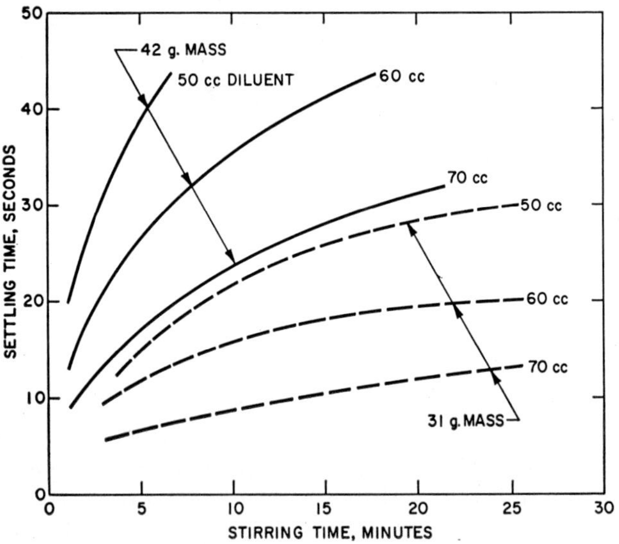

Figure 15.3. Curves plotted on a fine grid and on an open grid. Fine grids are needed only for interpolation or calibration.

using both an illustration and a detailed description or tabulation of the same data.

SELLING EDITORIAL SERVICES TO AUTHORS

If, in spite of such efforts to promote the preparation of good illustrations, an editor still receives poorly prepared figures, he must exercise tact and good judgment in attempting to remedy the situation without antagonizing the author. Often the editor has no budget for drafting or photographic services, and charges for revisions must be made directly to the research funds for the project. Then the editor has the even more difficult task of persuading the technical man that the errors are sufficiently serious to justify the expenditure of more time and money to correct them. At all times, the editor must balance the significance of an error against the time and money required to correct it.

The editor must be a person of some tact to be able to persuade an author to agree to making changes and to avoid injuring the author's pride. The correction of an error is such a delicate problem that an editor must continually strive so to educate technical authors that their original figures are free from mistakes. In the preparation of good illustrations it is indeed true that it is better to prevent errors than to correct them. "An ounce of prevention is worth a pound of cure."

SELECTED BIBLIOGRAPHY

1. American Standards Association, "Illustrations for Publication and Projection," Standard Y15.1, New York, 1958.

2. Cohn, Ernst M., "Editing Graphs for Publication," *Am. Doc.*, **6,** No. 2, 77-86 (1955).

3. Drake, Russell L., "A Medical Illustrator Looks at Medical Writing," *Miss. Valley Med. J.*, **76,** 37-40 (1954).

4. Jones, B. A., "Illustrating, Duplicating, and Binding the Technical Report," Chapter Nine of "The Technical Report," B. H. Weil (Ed.), New York, Reinhold Publishing Corporation, 1954.

5. Katz, Hyman H., "Technical Sketching and Visualization for Engineers," New York, Macmillan Company, 1949.

6. Kerekes, Frank, and Winfrey, Robley, "Report Preparation," Ames, Iowa, Iowa State College Press, 1951.

7. Levens, A. S., "Graphics in Engineering and Science," New York, John Wiley and Sons, Inc., 1954.

8. Michener, Thomas S., "Illustrating the Technical Presentation," *J. Chem. Educ.*, **31**, 318-23 (1954).

9. Singer, T. E. R. (Editor), "Information and Communication Practice in Industry," New York, Reinhold Publishing Corporation, 1958.

10. Staniland, L. N., "Principles of Line Illustration," Cambridge, Mass., Harvard University Press, 1953.

11. Ulman, Joseph N., New York, "Technical Reporting," Henry Holt and Company, 1952.

12. Waldo, W. H., "Instructing the Report Author," *J. Chem. Educ.*, **31**, 525-27 (1954).

Chapter Sixteen

Editing Tabular Data

ETHALINE CORTELYOU
Literary Analyst, Armour Research Foundation
Chicago, Illinois

Since scientists and engineers specialize in collecting, interpreting, and using technical data, it might be assumed that they would also be specialists in the most precise methods of reporting their data—equations, graphs, and tables. This is seldom true, however; while the equations are excellent in most reports and papers containing them, the graphs and tables are usually the poorest parts of a rough draft.

The scientific or engineering student is taught grammar, punctuation, some spelling, and composition in his freshman English course—which, unfortunately, may be his only college instruction in English. In his several courses in mathematics, he learns how to handle equations, and a little about making graphs. He receives relatively little instruction in tabulation, however, although tables are the most common form for reporting data. Moreover, the type of tables he does learn to make in courses on statistics or mathematics is not the proper one for reports and papers, since mathematical tables are usually set up as relations of variables with no dimensions given.

Although technical writing is basically expository writing characterized by techniques for reporting technical data, most of the many volumes on technical writing contain very little information on graphs, and practically nothing on tables (Ulman's "Technical Reporting"[3] is an exception). Because so very little is available, a simple compilation of 14 principles of data tabulation published in 1954[1] has been reprinted in several journals and has finally been revised and expanded to form a chapter of a book on information practices[2].

Most style manuals neglect tabulation. However, the manuals of the Bureau of Mines, Ethyl Corporation Research Laboratories, and Armour Research Foundation are among those which have very good coverage.

STYLE RULES

In editing tables, as in editing illustrations, "An ounce of prevention is worth a pound of cure." Time and effort spent in setting up a detailed style guide for authors to use in tabulating data will save many hours of editing, and will be a decided aid to good author-editor-typist relations. By setting rules for many of the mechanical details of tabulation, the editor eliminates the need for decisions on routine matters and reduces the possibilities of conflicts.

An editor must be quite firm in enforcing those rules of table format that affect only the appearance of the tables and not the interpretation of data. Although the appearance of tables may be a matter of personal preference on the part of the editor, the style guide should be followed by the author in order to insure uniformity of appearance. An author should have more important matters to concern himself with than whether the table number should be centered over the title or typed on the same line.

Such rules as the following are simply style rules intended for the typist and to assure uniformity of format:

1. The word "Table" and the number should be centered and underlined above the title of a table.

2. When a table is typed in the normal manner on a page, it should be centered between the left and right margins. When the table is typed the long way of the page, it should be centered between the normal top and bottom margins of the page.

3. The title of a table also should be centered, and it should be typed in all capitals. When the title is long and must be typed on two lines (single-spaced), the first line should contain a complete thought supplemented by the words typed on the second line. It may look best to type even a short title in two lines when the table is very narrow.

4. A double underscore is used, one-half space below the title, and is only as wide as the widest part of the table, whether that is the title or the columns. The scores are only one-half space apart.

5. A column heading should be centered over the column, and the underscore should be as wide as the widest part of the column, whether it is the heading or an item of data.

6. A column heading should be typed in upper and lower case letters. The unit of measurement should be set off by a comma rather than by parentheses.

7. From two to five spaces should be allowed between columns. A compact, but not crowded, table is easier to read and looks better than one sprawled across the page.

8. When a heading applies to two or more columns, it should be centered over them, and the underscore should be as wide as all the columns concerned.

9. The first line of the column heads should be one and one-half spaces below the double underscore.

10. All final underscores for column heads should be on

the same line, even though the heads may vary from one to three lines.

11. Two- and three-line heads should be single-spaced.

12. For ease in reading, double spacing is usually used in tables, but one and one-half spaces may be used for a long table. When data are presented in groups of two to four horizontal lines, single spacing may be used within a group and double spacing between groups.

13. When all numerical items in a column have the same dimensions (ft, sec, cps, etc.), they should be aligned by the decimal point when decimal fractions are involved, or by the last digit when they are whole numbers.

14. When the numerical items in a column have different units, each item should be centered in the column, or the column should be typed with a flush left margin. However, when several grouped items within such a varied column have the same units, those items should be aligned, as in 13.

15. A decimal fraction less than 1 must have a 0 before the decimal, except that only the first and last items of an entire column or group of such decimal fractions need to be set up in this manner.

16. Nonnumerical data (descriptions, colors, etc.) should be aligned at the left side of the column.

17. A run-in subtitle should be typed in upper and lower case letters. It may be centered or typed flush with the left margin. It should be underlined.

18. Sometimes the first column of a table contains entries descriptive of data in the rest of the table. Such a column is called a stub. When an entry in the stub is a group heading for two or more other entries, the group entry is typed at the left margin, and the other entries are indented below it. Extra space should be used to set off such a grouping.

19. A table is finished off with a double underscore, one-half space below the last line of data or type, and the same width and distance apart as the double score at the top of the table.

20. If the table has footnotes, however, the tabular part is separated from them by a single score one-half space below the last line of data and the same width as the double score at the top of the table. After the footnotes, the table is finished off with a double underscore, as in 19.

21. The first footnote is typed one and one-half spaces below the bottom score.

22. The use of the asterisk (*), the apostrophe ('), and quotes (") are preferable to either superscript numerals or letters for numbering footnotes.

23. Within a footnote, single spacing is used, but double spacing is used between footnotes.

CHECKLIST

In addition to the strictly style rules that must be enforced to insure tables that are uniform and attractive, there are other rules that must be enforced to insure tables that are understandable. Since cooperation from the authors rather than compliance by the typists is concerned, an editor might set up such rules as the following as a checklist for authors:

1. Every table must have both a title and a number.
2. Columns and rows must have identifying headings.
3. Units of measurement must be given.
4. The same units must be used for comparable properties or dimensions.
5. When possible, all items in a column should have the same unit, and the data must be aligned in the column by the decimal point.
6. When a table is set up with different units in one column, the data must be centered in the column or aligned at the left margin of the column, and must never be aligned by the decimal.
7. Footnote references must be "numbered" in sequence,

line by line, from left to right, across the entire table, except that each column or row heading should be considered to be a single line.

8. Similar tables must be set up in the same manner within a given report.

9. Ditto marks must not be used to indicate repetition in a table.

10. A run-in subtitle which applies only to the first part of a table should be typed below the column heads, and another subtitle should indicate the recording of data not affected by the first run-in subtitle.

These may seem to be insultingly obvious principles, but, in preparing this checklist, the alignment features of Rules 5 and 6 had to be added. Previously, alignment of comparable data by the decimal point seemed so obvious to the author that it was listed in the original article[1] only by inference, and in the rules supplied to Foundation typists. Recently, it was admitted that most new Foundation authors had to be advised of this "custom," so it cannot be considered to be universally known and should be stated in tabulation rules.

SUGGESTIONS

There are probably a few more rules that an editor might rigidly enforce in order to insure uniformity and readability of tables. However, before a rule is added to a style guide or checklist, it should be considered carefully to determine that it belongs in such a guide or checklist and that it is not a factor that would influence interpretation of data. Any principle that might affect interpretation of a table must be offered as a suggestion and not as a rule. The following principles, for instance, should be offered as suggestions:

1. When four or more sets of data are given, they should be presented in tabular form.

2. Even qualitative or descriptive comparisons and contrasts are more easily understood in table form.

3. The units used in a table should all be of the same system, such as metric, English, or apothecary.

4. If an item is repeated several times in a table, it should probably be set up in a footnote, the table title, or in a column or row heading, to eliminate the need for some of the repetition.

5. Column and row headings should be used to group related data.

6. In a table reporting a series of experiments or tests with some factors constant for certain groups and other factors variable, the data should be grouped according to the constant factors when possible.

7. The chronological order in which data were obtained is usually of little importance, and the sequence of data in the table should be determined by more significant factors.

8. Whenever possible, a table should be set up so that it can be typed on the page in the normal manner.

9. Data of only record interest should not be included in a table in the body of a report or paper.

10. Each type of data used in a table should be reported in a manner that will facilitate interpretation of all the data in the table.

11. Often, only a summary table should be presented in the main body of a report, and the supporting or record tables should be placed in an appendix.

12. The use of 10 to an exponent should be avoided whenever possible in tables.

Such principles should be offered only as suggestions because their observance may affect a reader's interpretation of the data. For instance, an author might object to Suggestion 1 because he might wish to mention several sets of data, but would not want to give them the significance of a numbered and titled table because he suspects that the data might be abnormal.

Table 16.1

FOG CLASSIFICATION FOR VARIOUS ADDITIVES*

Round No.	*Additive,* g*		*RH, %*	*Propellant, g*	*Fog Formation*
2426	A	14.1	97	200	Light
2427	B	13.2	97	200	Dense
2428	C	12.0	97	200	Dense
2429	D	15.7	97	200	Moderate
2430	E	8.1	97	200	Minimum
2431	F	5.4	97	200	Dense
2432	G	3.6	97	200	Moderate
2433	H	14.9	97	200	Moderate
2434	I	3.8	97	200	Dense
2435	J	14.1	97	200	Light
2451	A	28.2	95	200	Light
2452	B	26.3	95	200	Dense
2453	C	24.1	95	200	Dense
2454	D	31.5	95	200	Moderate
2455	F	10.7	95	200	Dense
2456	G	7.3	89	175	Light
2457	H	30.0	89	175	Light
2458	I	7.5	89	175	Dense
2459	J	28.1	89	175	Minimum
2462	A	28.1	97	175	Light
2463	B	26.4	97	175	Moderate
2464	C	24.0	97	175	Dense
2465	D	31.4	97	175	Moderate
2466	E	8.1	97	175	Minimum
2467	F	10.8	97	175	Dense
2468	G	7.2	97	175	Moderate
2469	H	29.8	97	175	Light
2470	I	7.6	97	175	Dense
2471	J	28.2	97	175	Light
2472	A	42.3	92	175	Light
2473	B	39.6	92	175	Moderate
2474	C	36.1	92	175	Moderate
2475	D	47.2	92	175	Light
2476	E	4.1	92	175	Minimum
2477	F	16.2	92	175	Dense
2478	G	10.8	92	175	Light
2479	H	44.7	92	175	Light
2480	I	11.4	92	175	Dense
2481	J	7.0	92	175	Minimum

* Additive code in Memorandum 11/27/57. The quantities of additives represent approximately 5, 10, 20, and 30% of a gram molecular weight.

Table 16.2

FOG CLASSIFICATION FOR VARIOUS ADDITIVES*

Fog Formation	*Round No.*	*Additive,* g*		*RH, %*	*Propellant, g*
Minimum	2430	E	8.1	97	200
	2459	J	28.1	89	175
	2466	E	8.1	97	175
	2476	E	4.1	92	175
	2481	J	7.0	92	175
Light	2426	A	14.1	97	200
	2435	J	14.1	97	200
	2451	A	28.2	95	200
	2456	G	7.3	89	175
	2457	H	30.0	89	175
	2462	A	28.1	97	175
	2469	H	29.8	97	175
	2471	J	28.2	97	175
	2472	A	42.3	92	175
	2475	D	47.2	92	175
	2478	G	10.8	92	175
	2479	H	44.7	92	175
Moderate	2429	D	15.7	97	200
	2432	G	3.6	97	200
	2433	H	14.9	97	200
	2454	D	31.5	95	200
	2463	B	26.4	97	175
	2465	D	31.4	97	175
	2468	G	7.2	97	175
	2473	B	39.6	92	175
	2474	C	36.1	92	175
Dense	2427	B	13.2	97	200
	2428	C	12.0	97	200
	2431	F	5.4	97	200
	2434	I	3.8	97	200
	2452	B	26.3	95	200
	2453	C	24.1	95	200
	2455	F	10.7	95	200
	2458	I	7.5	89	175
	2464	C	24.0	97	175
	2467	F	10.8	97	175
	2470	I	7.6	97	175
	2477	F	16.2	92	175
	2480	I	11.4	92	175

* Additive code in Memorandum 11/27/57. Quantities of additives represent approximately 5, 10, 20, and 30% of a gram molecular weight.

For Suggestion 11 the author of a table would have to decide which factors are constant and whether such constancy is significant for grouping.

Table 16.3

FOG CLASSIFICATION* FOR VARIOUS ADDITIVES

Propellant, g	200		175			
Rel Humidity, %	97**	95**	89′	97′	92′	
Additive, gmw %″	10	20	20	20	5	30
A	Light	Light		Light		Light
B	Dense	Dense		Mod		Mod
C	Dense	Dense		Dense		Mod
D	Mod	Mod		Mod		Light
E	Min			Min	Min	
F	Dense	Dense		Dense		Dense
G	Mod		Light	Mod		Light
H	Mod		Light	Light		Light
I	Dense		Dense	Dense		Dense
J	Light		Min	Light	Min	

* Min(imum), Light, Mod(erate), and Dense.

** Round No. in alphabetical sequence of additives:
97%: 2426-2435 95%: 2451-2455.

′ Round No. in alphabetical sequence of additives:
89%: 2456-2459 97%: 2462-2471 92%: 2472-2481.

″ Approximate gram molecular weight of each additive (code in Memorandum 11/27/57):

A: 141 B: 132 C: 120
D: 157 E: 81 F: 54
G: 36 H: 149 I: 38
J: 141

Exact listing of not only the rules to be enforced, such as the typist's rules and the checklist, but also the suggestions to authors will result in an improved quality of tables. Presenting examples of good tables is also a help to both typists and authors. Especially useful would be examples of the various correct ways the same set of data could be tabulated; examples might be selected to demonstrate how different sequences of data (or columns) might result in different evaluations or interpretations, as in Tables 16.1 to 16.4.

Table 16.4

FOG CLASSIFICATION* FOR VARIOUS ADDITIVES

Additive, gmw %**	30	20			10	5
RH, %	92′	89′	95′	97′	97″	92″
Propellant, g	175	175	200	175	200	175
A	Light		Light	Light	Light	
B	Mod		Dense	Mod	Dense	
C	Mod		Dense	Dense	Dense	
D	Light		Mod	Mod	Mod	
E				Min	Min	Min
F	Dense		Dense	Dense	Dense	
G	Light	Light		Mod	Mod	
H	Light	Light		Light	Mod	
I	Dense	Dense		Dense	Dense	
J		Min		Light	Light	Min

* Min(imum), Light, Mod(erate), and Dense.

** Approximate gram molecular weight of each additive (code in Memorandum 11/27/57):

A: 141 B: 132 C: 120 D: 157 E: 81
F: 54 G: 36 H: 149 I: 38 J: 141.

′ Round No. in alphabetical sequence of additives:

92: 2472-2481 89: 2456-2459 95: 2451-2455
97: 2462-2471

″ Round No. in alphabetical sequence of each additive (code in Memorandum 11/27/57):

97: 2426-2435.

EDITING TECHNIQUES

Even with such a "positive" preventive program, an editor may still receive poor tables. This is unfortunate because the editing of tables is certainly not a simple matter, and an editor must use the utmost tact in pointing out anything but style errors.

An author expects his editor to check punctuation and spelling; indeed, that is all the editing some authors are willing to accept. Freshman English may have been such a traumatic experience for some engineers, however, that they may welcome editing of text, especially with respect to tense and number of verbs. But a technical writer has an almost proprietary attitude toward his data, and will usually question the right of anyone to tell him how to handle them.

Unless the editor has a degree and some experience in a scientific or engineering field, he will find it difficult to get his technical authors to accept anything but the barest minimum of style editing of their tables. Any editing beyond style editing should therefore be offered as suggestions, no matter how technically competent the editor may be. However, time and patience are eventually rewarded, so that an editor can raise his standards by slow stages.

If the editor can consult an author while he is editing his tables, much better results can be obtained. Style editing can be done directly on the author's draft, but it is best to take the time to type a rough draft of any suggested rearrangement. An editor should remember that tables are not only difficult to design, but are also difficult to copy. Few typists or operators of monotype machines have memorized the positions of the number keys, so copying tables usually is a three-way switch for the eyes—from the draft to the keyboard, to the copy, and back to the draft. Therefore, the draft from which the final copy is to be made should be relatively clean. Moving a column of figures from the left side of a table to the right, or to the center, may be a vast improvement, but if any mistakes are made in copying the editor will usually be blamed, even though the author may have approved the change. To avoid errors in copying when a transposition is needed, a wise editor will use scissors and paste pot, or will even retype the rough draft. Because the data are considered "the meat" of a report or paper, an editor should think twice, and then twice again, before making any editing marks, and should carefully proofread any retyped portions.

Tables are difficult for an editor to edit, but they are such an important part of a technical report or paper that they definitely must be edited. Really, such editing is one of the most rewarding of editorial tasks, because any improvement in its tables is an obvious improvement of a report or paper.

REFERENCES

1. Cortelyou, Ethaline, "Some Fundamentals of Designing Tables of Data," *J. Chem. Educ.*, **31**, 590-93 (1954), and *Am. Docu.*, **5**, No. 3, 155-59 (1954).

2. Singer, T. E. R. (Ed.), "Information and Communication Practice in Industry," New York, Reinhold Publishing Corporation, 1958.

3. Ulman, Joseph N., "Technical Reporting," New York, Henry Holt and Company, 1952.

4. Walker, Helen M., and Walter N. Dursot, "Statistical Tables, Their Structure and Use," New York, Columbia University, 1936.

5. Weil, B. H. (Ed.), "The Technical Report," New York, Reinhold Publishing Corporation, 1954.

Part V

GENERAL

Chapter Seventeen

Editing Know-How—Techniques and Tools

I. H. JENKS and E. H. HUNTSMAN
Publications Division, Aluminium Laboratories Limited
Kingston, Ontario, Canada

The function of an editor is to make the author work harder and longer than he originally intended, says George Stevens.[91] In order to do this effectively for creditable publication (or for comprehension by readers), the editor, be he the editor of a journal, an industrial report, a technical manual, or a book, must have a large fund of knowledge of grammar, diction, punctuation, spelling, symbols, abbreviations, and special nomenclature, as well as the elements of printing and reproduction.

This knowledge, if not in the editor's head, must be easily available to him. The tools he uses are as important to him as are those of a carpenter or a surgeon if he, the editor, is to perform his task competently and thoroughly. Fortunately, these tools are numerous. They can be used to learn when, according to convention, he must capitalize the specific names of plants (but he must never capitalize the specific names of animals); how he can best indicate his suggestions for changes in a manuscript to an author; or when he is liable to charges of infringement of copyright.

ENGLISH USAGE

Dictionaries

By far the most numerous of the editor's tools are books on diction, grammar, style, and language. The choice is wide, and the field is broad. Webster's "New International" or "New Collegiate" dictionaries,[109, 108] and the "Shorter Oxford"[77] need only be mentioned as forming the backbone of any editorial library. More specific chemical and technical dictionaries such as Chambers',[100] Thorpe's,[96] Miall,[70] and the "Condensed Chemical Dictionary"[85] are all helpful for reference. In the *TWE Journal,* Walter[106] evaluates six technical dictionaries for the general user, although he does not discuss dictionaries for any particular field of science. Collins' "Authors' and Printers' Dictionary"[36] will also be helpful, as are many glossaries of printing terms found in Skillin,[89] Wiley,[111] the Chicago "Manual of Style,"[102] and Benbow.[19] Skillin[89] also includes a glossary of grammatical terms.

Grammar

Books on English grammar are legion; even those most useful to the technical editor are extremely numerous.[1, 2, 43, 48, 60, 61, 66, 74, 83, 94, 115] However, since the very fact that he is an editor means that he must have a more than average command of grammar, syntax, and diction, the books he consults most often will be those which are well-indexed and contain check-lists. With these he can quickly verify a point of grammar or settle an argument on punctuation.

Perrin's "Guide and Index to English"[81] is a text of this type. Not only are grammatical terms listed alphabetically and explained, but also difficult constructions (such as "compare with" versus "compare to") can be looked up quickly under the appropriate word. But the index, even with the accompanying instructions, will not suffice. Only on leafing through the book will the editor discover all that it contains. In fact, it is a question whether anyone would expect to find some of the entries such as "Heightened Style," "Good English," and "Written English." Nevertheless, once one becomes accustomed to its style, Perrin's "Guide" can be extremely useful.

Prepositions

More specific, but invaluable in its field, is Funk and Wagnalls' "Standard Handbook of Prepositions, Conjunctions, Relative Pronouns and Adverbs."[44] The long list of prepositional idioms (more than 2,000) shows the correct or permissible use of one or more prepositions with a given antecedent; that is, the antecedents are listed alphabetically, with all the permissible prepositions and illustrations of their use. For example, the editor will discover that he can not only live with, but he can also live at, live by, live down, live for, live in, live off, live on, live out, live to, and live within! This sort of book is most useful when editing manuscripts for an author whose native tongue is not English and thus is understandably confused by our idioms.

Compounding

An equally useful handbook is Ball's "Compounding and Hyphenation of English Words."[17] Here again is an alphabetical list, this time of compound and hyphenated words. Another list of two-noun phrases gives at a glance the preferred usage.

Synonyms

Further aids to the correct choice of words are, of course, dictionaries of synonyms, such as Roget's "Thesaurus,"[84] Fernald's "Handbook of Synonyms, Antonyms and Prepositions,"[41] and Webster's "Dictionary of Synonyms."[107] The section on "Use of Words" in Skillin's "Words into Type"[89] is useful, again because it includes alphabetical lists of such things as redundant phrases, trite expressions, appropriate expressions, words likely to be misused or confused, and American and British usage.

Spelling

Occasional difficulties in the spelling of technical words may be solved by consulting Hawley's "Technical Speller."[51] This book is particularly helpful with chemical nomenclature, although words in the fields of engineering, medicine, and physics are also included, as are nontechnical words "troublesome to spell and divide." "Catsup" (with the alternative "Ketchup"); the old spelling-bee sticker, "embarrass"; and even "cauliflower" presumably come in this category! Skillin,[89] the U. S. Government "Style Manual,"[103] and Mawson[62] also list words that are often misspelled.

Word Usage

Fowler's "Modern English Usage,"[42, 75] and Partridge's perhaps less well-known "Usage and Abusage,"[79] and "Dictionary of Clichés"[78] discuss in a useful and amusing way many points of grammar and word usage that are sometimes controversial. Partridge includes lists of clichés, similes, elegancies, and archaisms, all of which, he says, it is better to avoid. In addition to commonly serviceable items, Partridge includes usages that are little known. It may be surprising, and very possibly useful, to find (under the heading "Sports

Technicalities") that, when retiring to rest, a wolf "trains," a marten "trees," and a boar "couches." And not only that, but if you wish to be technically accurate in describing a group of birds, animals, fishes, or insects, you can find the "nouns of assemblage or company" which include a "charm" of finches, a "deceit" of lapwings, a "business" of ferrets, and a "troop" of dogfish!

Punctuation

Punctuation is the subject of many books and pamphlets.[25, 32, 33, 80, 93] Perhaps as useful an exposition as any for ready reference is that given in "Webster's New Collegate Dictionary."[108] However, punctuation is covered in detail in most style manuals[89, 102] as well as in many grammars and texts.

MARKING THE MANUSCRIPT AND PROOFREADING

Having found errors in grammar, punctuation, or spelling, the editor must indicate to the author his suggestions for improving the manuscript. The standard proofreaders' symbols are most commonly used in marking manuscripts, although in internal editing it is frequently sufficient to make corrections within the line, omitting marginal interpretation of marks. The so-called "standard proofreader's symbols," which are to be found in most style manuals,[24, 27, 38, 50, 57, 63, 68, 89, 102, 111] vary slightly in form and interpretation, but long custom in this country has fully established their meaning. In Europe, these correction marks have grown from custom into law. Standards have been set up in Holland and in Germany,[82] as well as in Great Britain.[24]

Trelease[97] gives detailed suggestions for making corrections in typed manuscripts, for insertions and transpositions, and for renumbering pages, while Baker[15] gives clear instructions for marking the typeset galley proofs and checking page proofs. The sharp distinction between proofreading and reading proof is pointed out by Gill.[47] These two processes are quite different operations, require a different approach and a different technique, and have a different objective. Proofreading is for the detection of errors in the transcription of manuscript to type, while reading proof is for the detection of errors in fact, punctuation, spelling, construction, consistency, and all related matters. Whether the author is the best person to correct the first proof is questioned by Poortenaar.[82] When the printer sends him the proof, he re-reads the text (so much easier to grasp, now that it is in print), and "is so pleased to see his offspring in print that he reads quietly on, with a gentle smile on his lips." But when the printer sends him the second proofs "he will be astonished to see what he had not seen before. For now the days of love at first sight are over, and he will compare those stupid printer's errors with his galley proofs—only to find that the mistakes are his own."

Almost every conceivable aspect of proofreading will be found in Lasky's "Proofreading and Copy Preparation."[57] What is not there will most certainly be in one or other of the 233 items in its general bibliography. The Chicago "Manual of Style,"[102] "Words into Type,"[89] and "Manuscript and Proof"[19] also deal in considerable detail with manuscript and proof correcting.

MECHANICS

Style Manuals

The mechanical details relating to manuscripts and printed matter, such as proper width of margins, form of captions, illustrations, etc., are the concern of the editor, too, since it is his responsibility to see that the style of the house or firm he represents is maintained in the format as well as in the content of its publications. In order to achieve this consistency in terminology, symbols, abbreviations, etc., almost every technical society, industrial firm, and technical book publisher has prepared a style manual for the guidance of its authors. Many of these style sheets are available, from the U. S. Government "Style Manual,"[103] the Chicago "Manual of Style,"[102] and the "Canadian Government Editorial Style Manual,"[28] to the pamphlets or sheets provided by most scientific societies[3, 4, 6, 7, 9, 10, 59] as guides to authors submitting papers to their journals. Typical of these latter, but more comprehensive than most, is the "Wistar Institute Style Brief,"[114] which devotes a large section of the manual to directions for the preparation of illustrations.

Illustrations

In addition to style manuals, a number of good texts are available on report writing,[49, 54, 99, 101, 110, 112] and several specific reference works[22, 31, 34, 35, 56, 87] offer many suggestions on format and the preparation of illustrations. The modern facilities for fine reproduction of photographs, drawings, and graphs are far removed from the facilities available to the illustrators of early scientific books, yet some of their devices were ingenious (if not now accepted practice). Sir Isaac Newton, according to Andrade,[11] adopted the device, in his book on "Opticks," of printing the diagrams on a folding plate with a blank space, the size of a page, nearest to the spine or attached edge, to make the diagrams still visible when the plate is extended but the book closed or a page turned over. Descartes solved the same problem by printing, for the reader's convenience, a diagram on *every* page on which it was mentioned—in his "Principia," one diagram is printed no less than 11 times!

Editing Standards

An interesting suggestion for a book of "Editing Standards" is proposed by Williams.[112] In it would be given a guide to, and a commentary on, the various authorities, and a list of new and interim usages. Here would be included technical terms and symbols commonly used by authors but not yet adopted by standardizing bodies. As a companion to this volume, Williams suggests a sample book of type sizes and layout—which might be entitled "Typographical Standards"—showing the main headings of papers and other articles, the various kinds of subheadings, and the manner of setting tabular data, giving footnotes, and indenting quotations.

Abbreviations

The use and form of abbreviations still vary, despite the efforts of many standardizing bodies, so that the choice of authority is arbitrary. A fairly comprehensive list of technical abbreviations and symbols is given by Zimmerman.[116] Other lists are to be found in the U. S. Government "Style Manual,"[103] Skillin's "Words into Type,"[89] British Standard 1991,[23] CSA,[29] ASA,[8] and the International Critical Tables.[104] Skillin also includes useful lists of standard French, German, Italian, and Spanish abbreviations.

A more specific but extremely useful book is Buttress' "World List of Abbreviations of Scientific, Technical and Commercial Organizations,"[26] while the standard abbreviations of the titles of scientific periodicals have been published by the ACS.[5] Suggestions for symbols for units of measurement which may be incorporated in physical equations are given in an issue of *Science*,[13] while Baker[15] reproduces a useful list of detailed recommendations by the London Mathematical Society for the guidance of mathematicians (or other scientists using equations) in preparing papers for the press. These are suggestions for transcribing symbols, easy to write in manuscript, into acceptable and available printers' fonts.

Choice of Type

Although it is usually the task of the publisher to choose the type face used for a technical publication, as well as such other details as size of type, dimensions of the type, dimensions of the type page, margins, and type of paper, the editor should at least understand the principles involved.[76] Many of these details can be found in the Chicago Manual,[102] Skillin's "Words into Type,"[89] Baker's "Technical Publications,"[15] and Melcher and Larrick's "Printing and Promotion Handbook."[69] The requirements of technical publications vary, of course, and the choice of type may depend not only on its suitability to the document, but also on its legibility—even its ability to remain legible through long exposure to the dirt of a workshop or plant.

Technical publications are produced for the sake of the matter contained in them, not primarily as works of art, and they must therefore be easy to read; the type must not hamper the reader. Aldous Huxley[20] tells of an eccentric typographical reformer who believes that we read too much and too easily, and thus do not take in the full significance of what we are reading. He therefore designed a set of letters so strange, and so unlike those we are used to, that our eyes would not be able to slide easily over the words, but would have to stop and decipher each one separately. In this way, we would have time to "suck out" the whole significance of the passage. Fortunately, we do not encounter many such!

Incidentally, internal editors must often work with special typewriters and phototypes ("cold type") and with "office" printing methods ("duplicating") in the production of technical reports and other documents which do not use typeset composition. A description of some of the equipment available, and hence the choice of "cold types," may be found in Weil's "The Technical Report."[110]

Size of Type

The size of type suitable for technical publications also varies, depending on the kind of document, its size, and the conditions under which it will be read. Small type permits more to be printed in a limited space, but small type on a large page should be set in two or more columns. Consideration of a page of text must include not only the proportions of the type-page and its relation to the proportions of the paper, but also the relation of the "tone" of the type to the white areas of the margins. It will be obvious that different margining will be necessary for a rectangle of black type with little space between lines, and for a rectangle of the same shape and size printed from light-face type with more space between the lines. In practice, few pages will be set in solid type. Norton[76] has some helpful remarks on these subjects.

Paragraph indentations break the text and make it a great deal more pleasing to the eye. Stephen Leacock,[58] the Canadian humorist, expresses it this way: "People are attracted to neat pages, artistically broken into trim sections. Set a thing into solid, unending blocks of type, into pages that never break, and few books could get over. Milton's 'Paradise' would be lost, and Dante's 'Inferno' would look like Hell."

The size of "type" is also important to the editor who publishes "typed" reports and similar documents. On one special typewriter, type fonts of various sizes and styles can be used interchangeably to give a printed appearance. Similar results for headings (large types) can be obtained by using photo-typesetters and similar devices. Electric and manual typewriters can be obtained with types of various sizes. Photo-reduction of typed layouts can also be employed. These are described in Weil's "The Technical Report."[110]

SCIENTIFIC WRITING

Finally, there are a number of books which the technical editor will certainly not use as tools but which will be stimulating and worthwhile additions to his library, since a study of scientific literature, and its growth, can broaden his outlook and sharpen his perception. Although he himself may not be a writer, familiarity with well-written scientific prose can help him to evaluate and improve the manuscripts he is responsible for editing. Emberger and Hall[40] have chosen many fine examples from a great variety of scientific papers and reports, while Drachman,[37] in his "Studies in the Literature of Natural Science," devotes 37 pages to a bibliography of sources of scientific literature. A few well-written historical engineering reports are given in full and discussed by Miller and Saidla in "Engineers as Writers."[73] The evolution of scientific English is a fascinating story told in a scholarly way by Andrews,[12] while the existence of the language of science is demonstrated in an informative and interesting manner by T. H. Savory.[86]

Other authors, of papers as well as books, have much to say that is related to editorial know-how. These include Arader,[14] Baker,[16] Banner,[18] Boyd,[21] Candelaria,[30] Elder,[39] Gates,[45] Gerard,[46] Hawley,[52] Hume-Rothery,[53] Keck,[55] McAllister,[64] McCartney,[65] McDaniel,[67] Michaelson,[71] Miller,[72] Smith,[90] Struck,[92] Thornton and Tully,[95] Trelease,[97] Tukey,[98] Waldo,[105] and Wilson.[113]

No editor can be expected to be sufficiently well-informed in all the fields of science which he encounters to detect errors obvious only to the specialist; on the contrary, the editor usually gives the author the benefit of the doubt, assuming that he will have treated accurately all matters relating to his own field of endeavor. Nevertheless, the more information the editor can acquire, or can have readily available, the greater will be the author's confidence in him and, therefore, the better will he be able to do his job of editing.

BIBLIOGRAPHY

This bibliography indicates the scope of tools available to the technical editor. Many similar references no doubt could be added. Whenever applicable, the inclusion of a bibliography has been noted, so that this list may be readily extended.

1. Adams, Henry H., "Techniques of Revision," New York, The Ronald Press, 1951.
2. Allbutt, Sir T. Clifford, "Notes on the Composition of Scientific Papers," London, MacMillan Company, 1925.
3. American Chemical Society, "Directions for Abstractors and Section Editors," Columbus, Ohio State University, 1952.
4. American Chemical Society, "Hints to Authors of Papers to be Presented at Meetings of the American Chemical Society," Washington, D. C., 1949.
5. American Chemical Society, "List of Periodicals Abstracted," *Chemical Abstracts,* **50,** 1J-314J (1956).
6. American Chemical Society, "Suggestions to Authors of Review Articles," *Chem. Reviews,* **48,** No. 2, 1-5 (1951).
7. American Institute of Chemical Engineers, "Guide to Authors and Guide to Speakers," New York.
8. American Standards Association, "Abbreviations for Scientific and Engineering Terms," New York, ASAZ10.1, 1941.
9. American Standards Association, "A Guide for Preparing Technical Illustrations for Publication and Projection," New York, ASA Y-15.1, 1958.
10. Anderson, J., and Thistle, M. W., "On Writing Scientific Papers," Canadian Journal of Research, Bulletin, 1947.
11. Andrade, E. N. da C., "The Presentation of Scientific Information," *Proc. Roy. Soc.* (*London*), **197,** No. A 1048, 1-17 (1949).

12. Andrews, Edmund, "A History of Scientific English. The Story of its Evolution Based on a Study of Biomedical Terminology," New York, Richard R. Smith, 1947.

13. Anon., "Symbols for Units of Measurement," *Science,* **120,** 1078-80 (1954).

14. Arader, Harry F., "A System for Testing and Increasing the Intelligibility of Technical Reports," *Science,* **121,** 537-39 (1955).

15. Baker, C., "Technical Publications, their Purpose, Preparation and Production," London, Chapman and Hall Ltd., 1955.

16. Baker, John R., "English Style in Scientific Papers," *Science,* **121,** 713-14 (1956).

17. Ball, Alice Morton, "The Compounding and Hyphenation of English Words," New York, Funk and Wagnalls, 1951.

18. Banner, E. H. W., "Technical Publicity. Some Notes on its Editorial Preparation," *Elec. Rev.,* **157,** 633-34, (1955).

19. Benbow, John, "Manuscript and Proof," New York, American Oxford University Press, 1937.

20. Bennett, Paul A., "Books and Printing," p. 348, Cleveland and New York, The World Publishing Company, 1951.

21. Boyd, W. C., "Arbitrary Editorial Changes in Scientific Papers," *Science,* **98,** 197-98 (1943).

22. British Standards Institution, "Bibliographical References," London, British Standard 1629, 1950.

23. British Standards Institution, "Letter Symbols, Signs and Abbreviations," London, British Standard 1991, Part 1, 1954.

24. British Standards Institution, "Table of Symbols for Printers' and Authors' Proof Corrections," London, British Standard 1219C, 1945.

25. Brittain, Robert, "Punctuation," Everyman's College Series, New York, Barnes & Noble, Inc., 1950.

26. Buttress, F. A., "World List of Abbreviations of Scientific, Technological and Commercial Organizations," London, Leonard Hill Limited, 1954.

27. Cambridge University Press, "Preparation of Manuscripts and Correction of Proof," Cambridge, England, 1951.

28. "Canadian Government Editorial Style Manual," Ottawa, The Queen's Printer, 1947.

29. Canadian Standards Association, "Standard Specification for Abbreviations for Scientific and Engineering Terms," Ottawa, C.S.A. Z85, 1943.

30. Candelaria, Nash, "Security and the Editor," *Science,* **121,** 528-30, 1955.

31. Carey, G. V., "Making an Index," Cambridge, England, Cambridge University Press, 1951.

32. Carey, G. V., "Mind the Stop," Cambridge, England, Cambridge University Press, 1948.

33. Carey, G. V., "Punctuation," Cambridge, England, Cambridge University Press, 1957.

34. Chaundy, T. W., Barnett, P. R., and Batey, Charles, "The Printing of Mathematics," London, Oxford University Press, 1954.

35. Cohn, Ernst M., "Editing Graphs for Publication," Paper presented before the American Documentation Institute, Washington, D. C., 1954.

36. Collins, F. H., "Authors' and Printers' Dictionary," Toronto, Oxford University Press, 1956.

37. Drachman, J. M., "Studies in the Literature of Natural Science," New York, Macmillan Company, 1930 (Bibliography).

38. Eidesheim, Julie, "The Editor at Work," New York, Farrar and Rinehart, Inc., 1929.

39. Elder, Joseph D., "Jargon—Good and Bad," *Science,* **119,** 536-38 (1954).

40. Embenger, M. R., and Hall, M. R., "Scientific Writing," New York, Harcourt Brace Co., 1955 (Bibliography).

41. Fernald, J. C., "Standard Handbook of Synonyms, Antonyms and Prepositions," New York, Funk and Wagnalls Company, 1947.

42. Fowler, H. W., "A Dictionary of Modern English Usage," 3rd Ed., London, Oxford University Press, 1926.

43. Fowler, H. W., and Fowler, F. G., "The King's English," 3rd Ed., London, Oxford University Press, 1930.

44. Funk and Wagnalls Editorial Staff, "Standard Handbook of Prepositions, Conjunctions, Relative Pronouns and Adverbs," New York, Funk and Wagnalls, 1953.

45. Gates, Marshall, "What the Editors of the Journal of the American Chemical Society Expect of Authors," *J. Chem. Educ.,* **31,** 456-58 (1954).

46. Gerard, R. W., "The Editing of Scientific Papers," *Science,* **106,** 289-90 (1947).

47. Gill, Robert S., "The Author Publisher Printer Complex," 2nd Ed., Baltimore, The Williams and Wilkins Company, 1949.

48. Gowers, Sir Ernest, "The Complete Plain Words," London, Her Majesty's Stationery Office, 1954 (Bibliography).

49. Guthrie, L. O., "Factual Communication," New York, Macmillan Company, 1948 (Bibliography).

50. Hart, Horace, "Rules for Compositors and Readers at the University Press, Oxford," London, Oxford University Press, 1954.

51. Hawley, G. G., "Technical Speller," New York, Reinhold Publishing Corp., 1955.

52. Hawley, G. G., "What the Technical Book Editor Expects of Authors," *J. Chem. Educ.,* **31,** 77-80 (1954). Also Chapter 5 in Waldo, W. H., "Better Report Writing," New York, Reinhold Publishing Corporation, 1958.

53. Hume-Rothery, W., "A Note on the Writing of Scientific Papers and Theses," *Research,* **10,** No. 5, 168-76 (1957).

54. Kapp, Reginald O., "The Presentation of Technical Information," London, Constable and Company Ltd., 1948.

55. Keck, Wendell M., "Manuscript Revision," *Chem. in Can.,* **8,** No. 12, 42-44 (1956).

56. Kinney, Mary R., "Bibliographical Style Manuals. A Guide to their Use in Documentation and Research," Chicago, Association of College and Reference Libraries, Monograph No. 8, 1953.

57. Lasky, Joseph, "Proofreading and Copy Preparation," New York, Mentor Press, 1954 (Bibliography).

58. Leacock, Stephen, "How to Write," New York, Dodd, Mead and Company, 1943.

59. London Mathematical Society, "Notes on the Preparation of Mathematical Papers," London, C. F. Hodgson and Son Ltd.

60. Masterson and Phillips, "Federal Prose," Chapel Hill, N. C., University of North Carolina, 1948.

61. Mathews, M. M., "Words: How to Know Them," New York, Henry Holt and Company, 1956.

62. Mawson, C. O. S., "Professional Book Editing," Boston, Mawson Editorial School, 1926.

63. Mawson, C. O. S., "Style-book for Writers and Editors," New York, T. Y. Crowell Company, 1926.

64. McAllister, D. T., "Is There Accepted Scientific Jargon?", *Science,* **121,** 530-32 (1955).

65. McCartney, Eugene S., "Does Writing Make an Exact Man?", *Science,* **119,** 525-28 (1954).

66. McCartney, Eugene S., "Recurrent Maladies in Scholarly Writing," Ann Arbor, University of Michigan Press, 1953.

67. McDaniel, W. B., "R. W. Gerard's Communication on the Subject of the Editing of Scientific Papers," *Science,* **106,** 491 (1947).

68. McGraw-Hill Book Company, Inc., "Preparing a Technical Manuscript," New York, 1953.

69. Melcher, Daniel, and Larrick, Nancy, "Printing and Promotion

Handbook," 2nd Ed., New York, McGraw-Hill Book Company, Inc., 1956.

70. Miall, Stephen, "A New Dictionary of Chemistry," London, Longmans, Green and Company, 1940.

71. Michaelson, H. B., "Semantics and Syntax in Technical Reports," *Chem. Eng. News,* **28,** No. 29, 2416-18 (1950).

72. Miller, Robert C., "The Care and Training of Authors," *Science,* **121,** 526-28 (1955).

73. Miller, W. J., and Saidla, L. E. A., "Engineers as Writers," New York, D. Van Nostrand Company Inc., 1953.

74. Montague, C. E., "A Writer's Notes on His Trade," London, Penguin Books, 1949.

75. Nicholson, Margaret, "American-English Usage, Based on Fowler's Modern English Usage," New York, Oxford University Press, 1957.

76. Norton, David M., "What the Technical Editor Should Know About Printing," Paper presented before the Soc. Tech. Writers and Editors, New York, 1957 (Bibliography).

77. Oxford University Press, "The Shorter Oxford English Dictionary," 2nd Ed., London, 1936.

78. Partridge, Eric, "Dictionary of Clichés," 4th Ed., London, Routledge and Kegan Paul, 1950.

79. Partridge, Eric, "Usage and Abusage," 4th Ed., London, Hamish Hamilton, Ltd., 1948.

80. Partridge, Eric, "You Have a Point There," London, Hamish Hamilton, Ltd., 1953 (Bibliography).

81. Perrin, Porter G., "Writer's Guide and Index to English," Chicago, Scott, Foresman and Company, 1950 (Bibliography).

82. Poortenaar, Jan, "The Art of the Book and its Illustration," London, George G. Harrap and Co. Ltd., 1935.

83. Quiller-Couch, Sir A., "On the Art of Writing," New York, G. P. Putnam's Sons, 1916.

84. Roget, P. M., "Thesaurus of English Words and Phrases," New York, Grossett and Dunlap, 1941.

85. Rose, Arthur and Elizabeth (Editors), "The Condensed Chemical Dictionary," 5th Ed., New York, Reinhold Publishing Corporation, 1956.

86. Savory, T. H., "The Language of Science," London, André Deutsch, 1953.

87. Schmid, C. F., "A Handbook of Graphic Presentation," New York, Ronald Press Company, 1954.

88. Shorb, Lura, and Beck, L. W., "Editorial Aids to Technical Writers," *Special Libraries,* **36,** 311-14 (1945).

89. Skillin, Marjorie E., and Gay, Robert M., "Word into Type," New York, Appleton-Century-Crofts Inc., 1948 (Bibliography).

90. Smith, Ralph B., "Publishing as Applied Science," *Science,* **119,** 538-39 (1954).

91. Stevens, George, "Author's Nursemaid," In: "Bookmaking and Kindred Amenities," pp. 45-56, New Brunswick, N. J., Rutgers University Press, 1942.

92. Struck, Herman R., "Recommended Diet for Padded Writing," *Science,* **119,** 522-25 (1954).

93. Sumney, George, "Modern Punctuation, Its Utilities and Conventions," New York, Oxford University Press, 1919.

94. Sypherd, W. O., "The Engineer's Manual of English," Revised Ed., New York, Scott, Foresman and Company, 1943 (Bibliography).

95. Thornton, J. L., and Tully, R. I. J., "Scientific Books, Libraries and Collectors," London, Library Association, 1954.

96. Thorpe, J. F., and Whiteley, M. A., "Thorpe's Dictionary of Applied Chemistry," 4th Ed., London, Longmans, Green and Company, 1937.

97. Trelease, S. F., "The Scientific Paper," Baltimore, Williams and Wilkins Company, 1947 (Bibliography).

98. Tukey, John W., "Systematization of Journal Practices," *Science,* **122,** 246 (1955).

99. Tuttle, R. E., and Brown, C. A., "Writing Useful Reports," New York, Appleton-Century-Crofts, Inc., 1956.

100. Tweney, C. F., "Chambers' Technical Dictionary," London, W. and L. Chambers, Ltd., 1940.

101. Ulman, Joseph N., "Technical Reporting," New York, Henry Holt and Company, 1952 (Bibliography).

102. University of Chicago Press, "A Manual of Style," 11th Ed., Chicago, 1949.

103. U.S. Government Printing Office, "Style Manual" (abridged), Washington, 1953.

104. U.S. National Research Council "International Critical Tables of Numerical Data, Physics, Chemistry and Technology," Vol. 1, pp. 16-17, New York, McGraw-Hill Book Company Inc., 1926.

105. Waldo, W. H., "Instructing the Report Author," *J. Chem. Educ.,* **31,** 525-27 (1954). Also "Better Report Writing," New York, Reinhold Publishing Corporation, 1958.

106. Walter, John A., "Technical Dictionaries," *TWE Journal,* 22-23 (Fall 1956).

107. "Webster's Dictionary of Synonyms," Springfield, G. and C. Merriam Company, 1942.

108. "Webster's New Collegiate Dictionary," Springfield, G. and C. Merriam Company, 1956.

109. "Webster's New International Dictionary," 2nd Ed., Springfield, G. and C. Merriam Company, 1942.

110. Weil, B. H., Editor, "The Technical Report," New York, Reinhold Publishing Corporation, 1954 (Bibliography).

111. Wiley, John, and Sons Inc., "Author's Guide for Preparing Manuscript and Handling Proof," New York, 1950.

112. Williams, G. E., "Technical Literature," London, George Allen and Unwin Ltd., 1948 (Bibliography).

113. Wilson, John H., "A Few Hints on Word Usage," *J. Chem. Ed.*, **33,** No. 11, 577-80 (1956).

114. Wistar Institute, "Style Brief," Philadelphia, Wistar Institute Press, 1934.

115. Witherspoon, A. M., "Common Errors in English and How to Avoid Them," New York, Barnes and Noble Inc., 1943.

116. Zimmerman, O. T., and Lavine, Irvin, "Scientific and Technical Abbreviations, Signs and Symbols," Dover, New York, Industrial Research Service, 1948.

Chapter Eighteen

Editing Translations

JULIAN F. SMITH
Lenoir Rhyne College
Hickory, North Carolina

and

T. E. R. SINGER
New York, New York

Technical translations, like other technical documents, usually need to be edited. Editing is costly, however, so the type and amount must be in accord with intended use. Thus fancy editing is not needed for translations made solely to convey information for local consumption.

Translations intended for wide circulation often need more intensive editing, however. Here, the prime consideration is wording the final result in good English. Fundamental requirements of a good translation include careful attention to shades of meaning; accurate rendering of technical terms into correct equivalents; appropriate use of trade terms or the jargon of the art concerned; and correct, appropriate use of abbreviations and symbols.

A fault to be shunned with special care is unexplained use of new technical terms, or of an abbreviation or symbol which may be unfamiliar to the readers. The editor must be certain that an explanation accompanies the first occurrence of each unfamiliar term, abbreviation, or symbol; later occurrences in the same translation may then stand unaided.

EDITING PATENT TRANSLATIONS

Editing translations of patents demands special care for at least three reasons:

1. Technical and legal terms are always bedfellows in patents, happily or not. The wording of patents is therefore directed to two audiences who differ in educational background. The translation of a patent must not lose any of the original comprehensibility to either audience, but it should not attempt to improve on the original in this respect.

2. Precision of statement is extremely important in any technical document, but in patents precision is not necessarily the original objective. Intentional vagueness is often built into patents, especially in their claims, and should therefore be retained in translations.

3. Patent translations are much used in Patent-Office hearings and in patent litigation. The outcome of such hearings or litigation often means the gain or loss of much money, or governs decisions for or against investing large sums. The translation should be neither more nor less informative than the original, so that its users can accept or reject facts on the same grounds and with the same confidence as if they were using the original patent.

Correcting apparent technical inaccuracies or misstatements in a patent specification is not a function of translators or editors, but sometimes an appended note calling attention to seeming defects may be desirable. This leaves the reader free to make his own decision. Again, neither translators nor editors are expected to apply merciful first-aid measures to the tortured involvements occurring in many foreign-language patents. Patents are not intended to be easy reading matter, even when first written in English (or what purports to be English). Labyrinthine thought tracks in a foreign language cannot be expected to come out as rectilinear thought tracks in English.

EDITING TRANSLATIONS OF TECHNICAL ARTICLES AND BOOKS

This same reasoning applies, although not quite so forcefully, in translating technical articles, books, and monographs. The language of origin will influence the result; lofty refinement of language and elegance of expression would not be economically justified even if they were desirable. Therefore, good technical translators neither seek nor attain that high literary polish which would conceal all traces of alien linguistic origin. However, a translation from technical French sounds more like original English than does one which has been laboriously hewn from solid German prose.

LANGUAGE FACTORS

Indeed, the familiar proverb about an impassable chasm between sericeous purses and porcine ears applies to translations and their editing. The tortuous, complicated processes of German thinking are faithfully represented by voluted, involuted, convoluted German paragraphs, winding in and out verb by verb and sentence by sentence. Thus, written expressions of German technical thought are seldom clear, and brevity is not to be expected. Translations of such matter, therefore, generally display a multiplicity of plural involvements. Even an experienced technical translator can hardly be called upon for pellucid English from murky foreign language copy at the going rates for translating. And any editor who tries to improve the output will soon find himself rewriting, not editing.

The translator will have moved verbs to positions that seem at least inoffensive to English-reading people; he will have reassembled detachable verb parts, restoring *zu* or *ab* from some far corner to *nehmen* in anglicizing to *in*crease or *de*crease. Such maneuvers alone will not suffice, however, if polished English is required.

Germans are fond of long, involved sentences, but so are writers in many other languages. The German specialty is a terminal concentration of verbs or bunched verb parts. Translators and editors find instances in which dichotomy or trichotomy of a sentence seems desirable, sometimes almost imperative. Such surgery may be feasible in making or editing literary translations, but it seldom is with technical matter, and practically never in patent specifications. When lucidity clashes with technical accuracy or the intended degree of precision, lucidity loses. No matter how strong the temptation to clarify, the original turbidity is to be retained.

When the editor decides that obscurity in a translation stems from accidental rather than intentional vagueness in the original, however, he should either edit accordingly, add an editorial note, or rewrite, according to circumstances. No firm rules can be laid down for dealing with these cases.

The French, rightly or wrongly, pride themselves on clear, logical writing. To us their *quatre-vingt-dix neuf* (four times twenty, ten, nine) seems an unnecessarily complicated way of saying *ninety-nine*, but it is probably no worse than Danish *halvfjerdsindstyve* (three and one half times twenty) for *seventy*.

In spite of numerical circumlocutions, however, French word order and sentence structure are not nearly as alien to English practice as are the customary arrangements of German words and sentences. The other Romance languages, too, are more like English than German in construction, although not always as straightforward as French.

Commonly, therefore, translations from French need less editing than those from German. Translators learn to eliminate such Gallicisms as *with of the acid* (avec de l'acide), and how to decide when *resumé* means a summary or when it means patent claims. They also learn to make allowance for the French tendency to be lenient about technical accuracy while relaxing none of their rigid insistence on writing good French.

Russian and some other languages, chiefly Slavic, have peculiar ways of modifying verb forms to indicate time, continuity, repetition, intensive action, etc. English lacks equivalent verb forms, and auxiliary verbs do not perform the same functions; proper rendition of the actual meaning is often impossible without circumlocutions. These cases need special editorial attention, lest they follow the wrong paths.

Chinese and Japanese, using ideographic characters instead of an alphabet, yield nearly incomprehensible English when literally translated. Native Chinese and Japanese translators tend to be too literal; this accounts for the oddities often found in their translations. Their output is difficult to edit, but often no more so than translations by native translators from European languages.

A source of error which crops up chiefly in Germanic and Romance languages is deceptive similarity between English and foreign terms. Thus, in German *Carbonsäure* is a general term for *carboxylic acid* and seldom means carbonic acid, while *Kohlensäure* does mean *carbonic acid* with no reference to acids from coal. Because French and its contributions to English have evolved along different paths, many English words have meanings different from the French parent. So *hydrate* is still good French for *hydroxide; prevenu* is participial *preceded,* not prevented; and *particulier* often means *private,* not particular. Editors must guard against survival of translation blunders from such sources.

All in all, readers are best served by translations based on idea units, not word units. Editors must be vigilant because excessive servility to words is among the prevalent weaknesses in translations.

Chapter Nineteen

Other Aspects

B. H. WEIL
Chief Editor, Technical Information Division
Esso Research and Engineering Company
Linden, New Jersey

Technical editing is really a diverse profession. This chapter will touch on the fundamentals of a few more phases.

EDITING TECHNICAL ADVERTISEMENTS AND PUBLICITY MATERIAL

Within organizations that employ advertising agencies, or have advertising departments that are not technically staffed, it often falls to the internal technical editor to "edit" technical advertisements, brochures, flyers, labels, etc., that have a technical message or slant. He does this editing from the viewpoint of management to insure the desired policy impact; from the technical viewpoint to insure accuracy, hence uphold the company's reputation;

and from the standpoint of the desired audience to help insure readability.

The internal technical editor is preferably involved in the management-level conferences which review plans for complete technical-advertising or technical-publicity programs. Here he gains insight on the intent of the programs, and may also contribute helpful ideas. He also comments on follow-up plans for the programs, and on those for specific advertisements or brochures.

Once subjects and slant have been determined, the technical editor is often given the assignment of channeling the necessary background information to the advertising agency's (or department's) writers, and of arranging conferences with his company's technical staff or management.

In editing drafts of the actual documents, the internal technical editor must exercise unusual caution against overediting and thus destroying creative freshness of approach; in fact, he must represent the agency's views to management and technical staff as well as their's to the agency. Many an advertisement or brochure has been lessened in impact because a technical director or engineer insisted on specific word changes instead of simply indicating that the present wording was not correct or desired, and why.

On the other hand, the internal technical editor must insist on technical accuracy, see that the slant is correct, and even *suggest* word changes when an impasse has been reached. He must develop management confidence in his judgment and agency confidence in his fairness and ability.

Internal technical editors sometimes turn writer and prepare brochures, advertisements, and the like. Or they may direct technical and/or nontechnical writers who do this work and also handle "technical" public relations. Aspects of this work have recently been touched on in a paper intended to interest students in technical communications as a profession.[2] The editor-turned-writer may also be interested in papers presented in a "Symposium on the Preparation of Technical Material for Development, Promotion, and Advertising."[1]

Incidentally, the editing of brochures, labels, and flyers often has legal as well as technical aspects. Knowledge of the legal factors involved may be a requirement of the technical editor assigned to review such documents.

EDITING TECHNICAL PROCEEDINGS AND COMPENDIA

The editor of the proceedings of a technical conference or symposium is ideally the chairman or arranger of the program. In this case he can early look beyond the oral presentation and plan also for publishing the papers involved—can organize the program to cover his subject adequately but without undue overlapping, and can also apprise authors of the mechanical needs for textual as well as oral versions.

If the editor of proceedings is simply assigned a group of papers to edit and unify, *ex post facto,* then his task will be more difficult. He will have to correspond more extensively with authors to obtain missing illustrations, and may find himself unable to remedy deficiencies in planning the program. Accordingly, if the chairman of a conference does not intend to edit its proceedings, he will do well to consult the editor who will do the work as long as possible before the program is presented.

Let us assume that the chairman has provided adequate instructions for oral versions—graphic aids included. What, then, should he or a separate editor require for the manuscripts needed for the proceedings?

First of all, the chairman should inform the author of the manuscript format required by the institution or publisher who will issue the proceedings. This may involve details such as margins, type of paper, size and spacing of typing, placing of internal headings (and whether they are to be capitalized and/or underlined), placing and format for tables and footnotes, bibliographic rules, size and style of drawings (including size of lettering and whether titles are also to be lettered or supplied separately), and size and type (glossy, usually) of photographs.

Secondly, the chairman or editor must establish deadlines, and must follow up with reminders. The experts who are the most desired speakers are usually very busy men, and some will put aside for tomorrow's attention what is not demanded today.

Once the drafts and illustrations are received, the editor can begin his work of reviewing for clarity, accuracy, and conciseness. If he is not the chairman of the conference, i.e., an expert in the subject, he may need to call for advice from those who are, and may have to relay their suggestions to the authors. More often, however, he is on his own and must suggest changes, ask questions, polish, etc., on his own initiative. From here on his role is much like that of the compiler of technical compendia.

A word about these latter editors. They are often dedicated individuals who see the need for books or whole series in specific fields, and who involve themselves personally in compiling the necessary texts. This they do by obtaining chapters or books from experts in phases of the subject, after (or sometimes even before) arranging with a publisher to issue the resultant book(s). Like the chairmen of symposia, they correspond extensively with the authors in order to clarify the intended coverage of specific chapters or volumes, and they notify authors of the necessary style requirements. Once the manuscripts are received they begin their careful review and analysis, and make their suggestions for changes.

The editors of technical compendia and proceedings are

subject to all the normal rules for working with professional authors. These are outlined in the first chapter of this book, and need little elaborating here. Within these bounds, however, the editor must insist on technical and factual accuracy, conciseness, and minimal overlapping of subject matter with other authors. In his zeal for sharp distinctions, however, he must not prevent each author from presenting a complete picture of his phase of the subject—must only insist that related phases be touched on rather than described in detail.

The editor must take pains not to suppress valid differences of opinion among his authors. In fact, he should call attention to these in his preface or in footnotes, so that the reader can gain a broader perspective. Unless his book or series is a dictionary or encyclopedia, where uniformity of approach is required, he should also preserve the authors' individual writing styles if these are clear and concise.

The editor of symposium or conference proceedings is usually an employee of the institution or society involved, or a professional volunteer. On the other hand, the editor of the technical compendium often does this as an independent professional. This means that he usually enters into a contractual relationship with his publisher—is either paid a fee for his work or receives a royalty based on sales. For writing their chapters, authors may share in this remuneration; if it is slight, however, their financial reward may be limited to complimentary copies of the book or their chapters.

In return for this fee or royalty the editor of a compendium supplies the publisher with edited chapters in the proper format, corrects galley proofs and page proofs, indexes the latter, and assists the publisher in promoting the sale of the book or series.

The editor of a symposium or conference proceedings often works directly with the printer; i.e., is his own publisher. Here he needs considerable knowledge of printing, including type styles, paper, and binding.

Editing technical proceedings or compendia, as mentioned, is somewhat like the editing of technical journals, but is also like the writing of technical books. There are times, indeed, when the editor of such documents almost wishes that he had elected to or been qualified to write instead of edit. But there is a personal challenge to working with fellow experts in creating a document that will be useful to many, and a degree of identification with the resulting work that is fully as rewarding as authorship.

EDITING TECHNICAL-PROCESS MANUALS AND HANDBOOKS

Samuel A. Miles, in Chapter Thirteen, has described the function of technical-equipment manuals and the principles of editing them. The type of equipment involved—missiles, computers, electronic equipment, etc.—is intricate, requires proper assembly, and must be operated in accordance with specified procedures.

In the chemical industry, however, one set of instructions may permit the erection of a plant to operate in a set way to yield a specific product, but it is practically unheard of for the operators of this plant to follow this procedure for long. Instead, they seek ways to increase yields, cut process time, substitute less expensive raw materials, and otherwise improve operations and products. All this requires research and development work, both before and after the plant is constructed. The results obtained, plus data from similar plants, can then be correlated in various ways to yield information on how one type of change will affect other variables.

Correlations of this type, combined with tables of physical

properties (of raw materials, products, and construction materials), specifications of component equipment, descriptions of procedures, and the hundred-and-one other bits of information useful in operating a chemical plant (or refinery, blast furnace, pulp mill, etc.) are known as technical-process manuals or "data books." They are compiled (edited) from many sources and many people, and represent the best judgment of the editor(s) and the technical management involved.

Technical-process manuals are usually first compiled by the engineering divisions of process-development companies or departments. Initially, they are based largely on pilot-plant data and on correlations developed by design engineers. Later, when the plant has been in operation for some time, or when data from other plants become available, revisions and additions become possible. These are made in the frame of reference for the subsequent use—of operating and maintenance engineers.

These process manuals are sometimes the collective responsibility of several engineers rather than one editor. On other occasions, however, their preparation is directed by a single editor who is versed in the technologies involved and in the writing and graphic-arts techniques required for proper presentation. Moreover, once a process manual has been issued, its revision will probably be the responsibility of a specific technical editor or editorial group.

It is obvious, of course, that here—as for the editing of the technical-equipment manual—the technical editor must be fully capable of preparing the correlations reported.

EDITING TECHNICAL SPECIFICATIONS

Many companies have installations that are widely separated, yet produce similar products or make parts that must be integrated with those made elsewhere. These products must often be virtually identical from plant to plant and time to time; otherwise, buyers will be unable to count on their quality and nature. The "parts" (including chemical intermediates) must also meet certain standards if the company itself is to operate efficiently.

To achieve these goals, companies employ standards—specifications—that they develop themselves or adapt from those prepared by authorities and groups. These specifications are usually written by those who develop them, but sooner or later an editor or editorial group takes over—develops uniform rules for their preparation, issues sets, distributes revisions, and spurs the development of new ones.

In editing these specifications, which are often thought of as relatively unglamorous documents, the experienced technical editor applies the practices and principles of editorship described in Chapter One. This is work with people and on useful technical documents. It affords excellent opportunities for constructive use of graphic arts and good writing and editing.

. . . AND SO ON

Even this chapter has not exhausted the subject of technical editing. For example, there remains the editing of technical patents, both by the patent administrator who reviews drafts prepared by his staff or by outside agents, and by the technical man who reviews patents describing his own or his group's inventions. Nor have we dealt with the desirable requirements for college courses (graduate-level, presumably) in technical editing. And there is much more that we could have discussed at the nuts-and-bolts level.

History has taught us, however, that every ship must have its launching, and that this launching would never take place if the builder insisted on installing every improvement as it is announced. We must therefore leave it to future editions of this book to report in some detail the many phases of technical editing which have been slighted or neglected. For this future effort, the suggestions of readers will be indeed welcome.

REFERENCES

1. Authors, "Abstracts of Papers Presented at Atlantic City, New Jersey, September 16-21, 1956, before the Division of Chemical Literature of the American Chemical Society," *Chem. Literature,* **8,** No. 3, 14-17 (1956).

2. Starr, C. E., Jr., Weil, B. H., and Bragg, B. L., III, "Career Opportunities in Technical Communications," *Chem. Eng. News,* **36,** No. 4 (2), 9-11 (1958).

Author Index

Adams, H. H., 244
Allbutt, T. C., 244
Anderson, J., and Thistle, M. W., 244
Andrade, E. N. da C., 239, 244
Andrews, E., 243, 245
Arader, H. F., 243, 245

Baker, C., 240, 241, 245
Baker, J. R., 243, 245
Ball, A. M., 235, 245
Banner, E. H. W., 243, 245
Beck, L. W., joint author. *See* Shorb, L., and Beck, L. W., 249
Benbow, J., 234, 245
Bennett, P. A., 245
Bixler, G. H., 133
Bland, W. F., 115
Bliss, A. D., 91
Boyd, W. C., 243, 245
Bragg, B. L., III, joint author. *See* Starr, C. E., Jr., Weil, B. H., and Bragg, B. L., III, 265
Breth, R. D., 50
Brittain, R., 245
Brown, C. A., joint author. *See* Tuttle, R. E., and Brown, C. A., 249
Bush, V., 13, 182
Buttress, F. A., 240, 245

Candelaria, N., 243, 245
Carey, G. V., 245, 246
Chaundy, T. W., 246
Cohn, E. M., 215, 246
Collins, D. N. and Jones, B. A., 187
Collins, F. H., 234, 246
Connelly, W. J., 202
Connolly, L. D., 80
Cortelyou, E., 16, 217, 229
Cortelyou, E., and Jones, B. A., 203
Crane, E. J., 79

Davis, D. S., 15, 40
Drachman, J. M., 243, 245
Drake, R. L., 215
Dursot, W. N., joint author. *See* Walker, H. M., and Dursot, W. N., 229

Eidesheim, J., 245
Elder, J. D., 243, 246
Embenger, M. R., and Hall, M. R., 243, 246

Fernald, J. C., 236, 246
Fleischer, M., and Hooker, M., 79
Ford, G., 79
Fowler, H. W., 236, 246
French, B. J., 31

Gates, M., 243, 246
Gerard, R. W., 243, 246
Gill, R. S., 246
Gowers, E., 246
Gunning, R., 40
Guthrie, L. O., 246

Hall, M. R., joint author. *See* Embenger, M. R., and Hall, M. R., 243, 246
Hamor, W. A., 165
Hart, H., 247
Hawley, G. G., 155, 236, 243, 247
Hooker, M., joint author. *See* Fleischer, M., and Hooker, M., 79
Hume-Rothery, W., 243, 247
Huntsman, E. H., joint author. *See* Jenks, I. H., and Huntsman, E. H., 233
Huxley, A., 241

Jackson, L., 79
Jenks, I. H., and Huntsman, E. H., 233
Jones, B. A., 202, 216
Jones, B. A., joint author. *See* Collins, D. N., and Jones, B. A., 187
Jones, B. A., joint author. *See* Cortelyou, E., and Jones, B. A., 203

Kapp, R. O., 247
Katz, H. H., 216
Keck, W. M., 243, 247
Kerekes, F., and Winfrey, R., 216
Kinney, M. R., 247
Kurt, O. E., joint author. *See* Weil, B. H., and Kurt, O. E., 87

Lane, J. C., 41
Lane, J. C., joint author. *See* Weil, B. H., and Lane, J. C., 15
Larrick, N., joint author. *See* Melcher, D., and Larrick, N., 241, 247
Lasky, J., 238, 247
Lavine, I., joint author. *See* Zimmerman, O. T., and Lavine, I., 250
Leacock, S., 242, 247
Levens, A. S., 216

McAllister, D. T., 243, 247
McCartney, E. S., 243, 247
McDaniel, W. B., 243, 247
Masterson and Phillips, 247
Mathews, M. M., 247
Mawson, C. O. S., 236, 247
Meites, L., 103
Melcher, D., and Larrick, N., 241, 247
Mento, M. A., joint author. *See* Weil, B. H., Schoengold, M. D., and Mento, M. A., 62
Mento, M. A., joint author. *See* Schoengold, M. D., Weil, B. H., and Mento, M. A., 62
Miall, S., 234, 248
Michaelson, H. B., 243, 248
Michener, T. S., 216
Miles, S. A., 167, 262
Miller, R. C., 16, 243, 248
Miller, W. J., and Saidla, L. E. A., 243, 248
Montague, C. E., 248

Nicholson, M., 248
Norton, D. M., 242, 248
Noyes, W. A., Jr., 105

Partridge, E., 236, 248
Perrin, P. G., 234, 248
Phillips, joint author. *See* Masterson, and Phillips, 247
Poland, L. B., joint author. *See* Weil, B. H., and Poland, L. B., 16
Poortenaar, J., 238, 248

Quiller-Couch, A., 248

Rathbone, R. R., 15
Reyling, P. M., 19
Roget, P. M., 236, 249

Rose, A., and E., 248
Rose, E., joint author. *See* Rose, A., and E., 248
Ross, G. K., 201

Saidla, L. E. A., joint author. *See* Miller, W. J., and Saidla, L. E. A., 243, 248
Salmons, N., 201
Samstag, N., 16
Savory, T. H., 248
Schmid, C. F., 248
Schoengold, M. D., 62
Schoengold, M. D., Weil, B. H., and Mento, M. A., 62
Schoengold, M. D., joint author. *See* Weil, B. H., Schoengold, M. D., and Mento, M. A., 62
Sewell, W., 79
Shidle, N. G., 39, 40
Shorb, L., and Beck, L. W., 249
Singer, T. E. R., 216, 229
Singer, T. E. R., joint author, *See* Smith, J. F., and Singer, T. E. R., 251
Skillin, M. E., and Gay, R. M,. 234, 236, 240, 249
Smith, J. F., and Singer T. E. R., 251
Smith, R. B., 243, 249
Southgate, M. T., 16
Staniland, L. N., 216
Starr, C. E., Jr., Weil, B. H., and Bragg, B. L., III, 265
Stephens, I. R., 19
Stevens, G., 233, 249
Struck, H. R., 243, 249
Sumney, G., 249
Sypherd, W. O., 249

Thistle, M. W., 16
Thistle, M. W., joint author. *See* Anderson, J., and Thistle, M. W., 244
Thornton, J. L., and Tully, R. I. J., 243, 249
Thorpe, J. F., and Whiteley, M. A., 234, 249
Trelease, S. F., 237, 243, 249
Tukey, J. W., 243, 249
Tully, R. I. J., joint author. *See* Thornton, J. L., and Tully, R. I. J., 243, 249
Tuttle, R. E., and Brown, C. A., 249
Tweney, C. F., 249

Ulman, J. N., 216, 229, 249

Van Antwerpen, F. J., 12, 13, 15

Waldo, W. H., 15, 216, 243, 249
Walker, H. M., and Dursot, W. N., 229
Wall, F. E., 16
Walter, J. A., 234, 249
Weil, B. H., 2, 15, 40, 229, 242, 243, 250, 257
Weil, B. H., and Kurt, O. E., 87
Weil, B. H., and Lane, J. C., 15
Weil, B. H., and Poland, L. B., 16
Weil, B. H., joint author. *See* Schoengold, M. D., Weil, B. H., and Mento, M. A., 62
Weil, B. H., Schoengold, M. D., and Mento, M. A., 62
Weil, B. H., joint author. *See* Starr, C. E., Jr., Weil, B. H., and Bragg, B. L., III, 265
Whiteley, M. A., joint author. *See* Thorpe, J. F., and Whiteley, M. A., 249
Williams, G. E., 240, 250
Wilson, J. H., 243, 250
Winfrey, R., joint author. *See* Kerekes, F., and Winfrey, R., 216
Witherspoon, A. M., 250

Zimmerman, O. T., and Lavine, I., 240, 250

Subject Index

Abbreviations, 240, 241
Abstracts, 62–79
  functions of editors of, 63–65
  on file cards, 77–79
  prepared by central services, 73, 74
  publishing bulletins of, 68–77
    arrangement in, 68–70
    coverage in, 73, 74
    distribution practices in, 76, 77
    duplicating in, 75, 76
    editorial comments in, 65, 72
    graphic-arts factors in, 75, 76
    highlighting of items in, 70–73
    section headings for, 69, 70
    selecting items for, 73–75
  readability of, 64, 66, 67
  reasons for internal preparation of, 62, 63
  simultaneous preparation of bulletins and file cards, 77–79
Advertisements, 257–59
API Technical Abstracts, 74
Armed Forces, 170
Armour Research Foundation, 206, 218
Articles. *See* Papers
Authors. *See* Writers

Battelle Memorial Institute, 206, 207
Books. *See also* Compendia; Editors, of books; Proceedings
  publisher editing of, 155–65
    evaluating manuscripts in, 160, 161
    following up on writers in, 159–61
    preparing manuscripts for printer in, 162–64
    procuring in, 157, 159
  status of editors of, 12
  translation of, 253
  utility of style manuals for, 7
  value of contracts for, 2
Brochures, 257–59

Chrysler Corporation, 201
Color, 124
  use in slides, 193, 194
Communication techniques, in technical-business journals, 116–25
Compendia, 259–62. *See also* Books, Proceedings
  arranging for, 260
  style rules for 259, 260
Compounding, 235
Consultants, as teachers of internal courses, 8, 9
Courses, on writing, 7–9

Data books. *See* Manuals
Data charts, 190
Department of Defense, 170

Dictionaries, 234
Documents, 6. *See also* specific types

Eastman Kodak Company, 201
Editors. *See also* Professional aspects
  as teachers of internal courses, 8, 9
  as writers, 181
  educational requirements for, 13, 14
  function in preparing slides, 189
  handling of news releases, 133–51
  internal
    management needs for, 81
    management obligations to, 82
    training of, 80–88
      cost of, 87
      indoctrination courses in, 83, 84
      in graphic arts, 86, 87
      supervision during, 84–86
  management roles of, 10–12
  of advertisements and publicity material, 257–59
  of books, 155–65. *See also* Books, publisher editing of
    executive, 156–59
    levels of, 156–64
    qualifications of, 155, 156, 159, 161, 164
    second-level, 159–61
    third level, 162–64
  of compendia, 259–62
  of illustrations for reports and papers, 205, 206, 215
  of proceedings, 259–62
  of process manuals, 263
  of specifications (for products), 264
  of technical-business journals. *See also* Journals, technical-business
    difference from editors of technical-society journals, 125–27
    practices of, 129–31
      approval of articles, 129, 130
      author's check of proofs, 130
      honorariums and reprints, 130, 131
  of technical-equipment manuals, 171–83. *See also* Manuals, technical-equipment
    difference from other editors, 167, 172, 173, 180–82
    functions of, 173–83
      checking, 180
      defining nomenclature specifications, 179
      defining style specifications, 179
      defining technical specifications, 173–79
      directing the proofing program, 180
      personalizing the job, 180, 181
      professionalizing the job, 181, 182
  of technical-society journals. *See also* Journals, technical-society
    action on reviews, 94, 95
    managing, 100, 101
    modern concept of, 115, 116
    qualifications for, 91–93
  of translations, 251–56
  relations with illustrators, 4, 5
    in preparing slides, 188, 189
  relations with printers, 5

Editors (*continued*)
relations with reviewers of technical-society papers, 98, 99, 112–14
relations with writers, 2–4
as regards illustrations, 215
day-to-day, 9, 10
during editorial training, 85, 86
in internal editing of papers, 38–40
in preparing slides, 188–90, 196–98
of abstracts, 63–65
of compendia chapters, 260, 261
of papers, 98, 99
as regards patent aspects, 51–61
of reports, 20–23
of tables, 227–29
role in internal editing of papers, 33–38
Educational aspects, 6–10. *See also* Courses, on writing; Editors, relations with writers; Style manuals
Engineering Index Service, 74
English usage. *See* Techniques and tools
Equations, 204, 217
Equipment manuals. *See* Manuals
Esso Research and Engineering Company
highlighting of abstracts, 71
section headings for abstracts bulletins, 69
style of abstracts, 66
Ethyl Corporation, 74, 201, 218

Flyers, 257–59

Grammar, 234
Graphic aids and other exhibits, 185–229. *See also* Graphic arts, Illustrations
illustrations for reports and papers, 203–16
slides and other graphic aids, 187–201
tabular data, 217–29
Graphic arts. *See also* Graphic aids, Illustrations
in design of abstracts bulletins, 75
in report publication, 27, 28
training internal editors in, 86, 87
Graphs. *See also* Illustrations
value of grids in, 213, 214

Headlines, 116–18

Illustrations, 217, 218. *See also* Graphic arts, Graphic aids, Slides
for reports and papers
editorial responsibility and control of, 205, 206
scheduling preparation of, 204, 205
style-manuals for, 206, 207
value of examples in, 207–15
in books, editing and scaling of, 162
in organization magazines and newsletters, 47
in technical-business journals, as part of modern presentation, 121–23
style manuals for, 206–15, 239
Illustrators
relations with editors and writers, 5
in preparing slides, 188, 189
Indoctrination courses, in training internal editors, 83, 84
Internal editing, 17–87
from the patent standpoint, 51–61
of abstracts, 62–79

Internal editing (*continued*)
of organization magazines, 41–50
of papers and articles, 31–40
of reports, 19–30

Journals, 89–151
modern communication techniques in. *See* Journals, technical-business
news releases in, 133–51
status of editors of, 12
style rules for, 37, 38
technical-business, 115–32. *See also* Editors, of technical-business journals
difference from technical-society journals, 125, 126
editorial practices for, 129–31
modern communications techniques in, 116–25
headlines in, 116–18
illustrations in, 121–23
"optional reading" in, 123, 124
organizing and structuring articles in, 119, 120
use of color and typography in, 124, 125
"variable-speed reading" in, 124, 125
sources of papers for, 127–29
"contributed," 128, 130, 131
meeting papers, 128, 129
staff written, 127, 128
technical-society. *See also* Editors, of technical society journals
declining papers for, 96–98
difference from technical-business journals, 125, 126
editor's viewpoint on, 91–102
managing-editor activities for, 100, 101
qualifications for editors of, 91–93
review by referees for, 93, 94, 103–14. *See also* Reviewing, for technical-society journals
editorial action on, 94, 95, 113
reviewer's viewpoint on, 103–14
utility of style manuals for, 7
value of policy statement for, 2

Knolls Atomic Power Laboratory, 207

Labels, 257–59
Legal aspects, 259. *See also* Patent considerations
Limits, 3

Magazines. *See* Journals, technical-business; Organization magazines and newsletters
Management
need for clarity in policy, 2, 4
needs
for editors, 81
for streamlined presentations, 6
obligations to editors, 82
recognition of writers needs, 7, 8
status of editors, 10–12, 82
viewpoint in editing advertisements, 257, 258
Manuals
process, 262, 263
technical-equipment, 167–83. *See also* Editors, of technical-equipment manuals

Manuals (*continued*)
cost of, 170, 171, 182, 183
function of, 168, 169, 171
scope of, 169
Military-equipment manuals. *See* Manuals, technical-equipment
Models and test pieces, 190
Movies, 190

News releases, 133–51
Newsletters. *See* Organization magazines and newsletters

"Optional reading," 123, 124
Organization magazines and newsletters, 41–50
functions of editors for, 48–50
periodicity of, 46
publishing of, 48–50
purpose as determining factor for, 41–48
effect on editing decisions, 46–48
ratio of features to news in, 46
ratio of illustrations to text, 47
typography in, 44–46

Papers
difference between technical-society and technical-business, 125, 126
for technical-business journals
contributed, 128
honorariums for, 130, 131
from meetings, 128, 129
staff-written, 127, 128
illustrations for, 203–16
internal editing of, 31–40
from the patent standpoint. *See* Patent considerations
reader's needs as factors in, 34, 35
role of editor in, 33–38
role of review committee in, 32, 33
to meet journal specifications, 37, 38
value of, 31, 32
reviewing of. *See* Reviewing, Journals, technical-society
tables for, 217–29
translations of, 253
Patents, 264
Patent considerations, 51–61
advisability of publication, 51
in method of data presentation, 55–57
need for consistency between statements in patents and publications, 53, 55–60
publication before filing applications, 53, 54
publication during pendency of applications, 52, 53
regarding novelty statements, 58, 59
regarding operability statements, 59–71
regarding translations, 252, 253
regarding utility statements, 57, 58
Petroleum Refining Developments, 74
Photographers. *See* Illustrators
Prepositions, 235
Proceedings, 259–62. *See also* Books, Compendia
advance plans for, 259, 260
style rules for, 259, 260
Process manuals. *See* Manuals, process
Professional aspects, 10–14, 164–66
as regards technical-equipment manuals, 172, 173, 180–83
educational requirements, 13, 14
in book editing, 164–66

Professional aspects (*continued*)
organizational status, 10–12. *See also* Management
for report editing, 25, 27
professional status, 12, 13
Proofreading, 237, 238
of technical-society journals, 99–101
Psychological aspects, 2–6. *See also* Editors, Illustrators, Readers, Writers
Publishing
of abstracts bulletins, 68–77
of books, 162–64
of compendia, 261, 262
of organization magazines and newsletters, 48–50
of proceedings, 261, 262
of tables, 217–29
of technical-business journals. *See* Journals, technical-business
of technical-society journals, role of editor in, 99
techniques and tools for, 233–50
type style and size for, 241–43
use of color and typography in, 124, 125
Punctuation, 237

Quality-control aspects, 30, 31, 183

Readers
capturing of interest in papers, by internal editing, 35–37
needs of, 5, 6
as a consideration in editing organization magazines and newsletters, 41–48
as a consideration in internal editing of papers, 34, 35
for individual copies of abstracts bulletins, 76
for information from news releases, 134, 135
for quick-reading abstracts bulletins, 65–73
of technical-business journals, methods of attracting attention of, 116–25
of technical-society journals, responsibilities of reviewers to, 109–11
use of "variable-speed reading" for, 124, 125
Reading proof, 237, 238
Reports, 19–30
editor-writer relationships in preparing, 20–23
illustrations for, 203–16
in large organizations, 26–28
in smaller organizations, 29
purpose of, 20
role of editorial supervisor for, 25
tables for, 217–29
type style and size for, 241–43
use of style manuals for, 23–25
Reviewers. *See* Reviewing, Journals, technical-society
Reviewing
for technical-society journals,
anonymity in, 94, 108, 109
criteria in, 93, 94
editor's viewpoint on, 93, 94
editorial action on reviews from, 94, 95, 113, 114
reports on, 94, 112–14
responsibilities of reviewers in, 104
to editors, 104–06
to readers, 109–11
to subject authorities, 111, 112
to writers, 106–09
value of, to writers, 95, 106–09
Rewriting, of news releases, 147–51

Slides, 187–202
as poster art, 194
color in, 193, 194
editor's role in preparing, 188, 189
examples of, 198–200
negative, 192, 193
positive, 191, 192
presentation of, 200, 201
scheduling preparation of, 194–98
utility of, 191
as notes for presentation, 200, 201
writer as boss in preparing, 188, 189
Specifications (for products), 263, 264
Spelling, 236
Style manuals, 2, 238, 239
for compendia, 259, 260
for illustrations, 206, 207, 239
value of examples in, 207–15
for proceedings, 259, 260
for reports, 21, 23–25
for tables, 218
for clarity, 221, 222
for readability, 222–27
for uniformity and attractiveness, 218–21
purposes of, 6, 7
Supervision
during training of internal editors, 84–86
of report editing
in large organizations, 26–28
in smaller organizations, 29
role of supervisor in, 25
Synonyms, 236

Tables, 203, 204, 217–29
examples of, 224–27
rules for, 218
for clarity, 221, 222
for readability, 222–27
for uniformity and attractiveness, 218–21
statistical, 217
techniques for handling, 228, 229
Technical-business magazines. *See* Journals, technical-business
Technical-society journals. *See* Journals, technical-society
Techniques and tools, 233–50
abbreviations, 240, 241
choice of type, 241, 242
English usage, 234–37
compounding, 235
dictionaries, 234
grammar, 234
prepositions, 235
punctuation, 237
spelling, 236
synonyms, 236
word usage, 236
for proofreading, 237, 238
for reading proof, 237, 238
size of type, 242, 243
standards for editing, 240
style manuals, 238, 239. *See also* Style manuals
for illustrations, 239
Television, 190, 191
Translations, 251–56
fundamental requirements of, 251, 252
language factors in, 253–56
of books, 253
of papers, 253
of patents, 252, 253
Type
choice of, 241, 242
size of, 242, 243
Typography, in organization magazines and newsletters, 44–46

"Variable-speed reading," 124, 125

Word usage, 236
Writers
  needs for additional training of, 7–9
  of books
    following up on, 159–61
    procurement of, 157, 159
  of technical-publicity material, 258
  of technical-society papers
    benefits of reviewers to, 95, 106–09
    relations with editors, 98, 99
    relations with reviewers, 98, 99, 113
  relations with editors
    as regards illustrations, 215
    as regards papers, 98, 99
    as regards reports, 20–23
    as regards tables, 227–29
    day-to-day, 9, 10
    in internal editing of papers, 38–40
    in preparing slides, 188–90, 196–98
    of compendia, 260, 261
    on patent aspects, 51–61
  value of occasional editing, 22
Writing, 1. *See also* Books, Editors, Journals, Manuals, Readers, Reports, Rewriting, Writers
  scientific, 243, 244